Mechanical

Discipline-Specific Review for the FE/EIT Exam

Second Edition

Michel Saad, PhD, PE
Abdie H. Tabrizi, PhD
with Michael R. Lindeburg, PE

Professional Publications, Inc. • Belmont, California

How to Locate and Report Errata for This Book

At Professional Publications, we do our best to bring you error-free books. But when errors do occur, we want to make sure you can view corrections and report any potential errors you find, so the errors cause as little confusion as possible.

A current list of known errata and other updates for this book is available on the PPI website at **www.ppi2pass.com/errata**. We update the errata page as often as necessary, so check in regularly. You will also find instructions for submitting suspected errata. We are grateful to every reader who takes the time to help us improve the quality of our books by pointing out an error.

MECHANICAL DISCIPLINE-SPECIFIC REVIEW FOR THE FE/EIT EXAM
Second Edition

Current printing of this edition: 1

Printing History

edition number	printing number	update
1	3	Minor corrections.
1	4	Updated front matter.
2	1	New edition. Copyright update.

Printed in the United States of America

Professional Publications, Inc.
1250 Fifth Avenue, Belmont, CA 94002
(650) 593-9119
www.ppi2pass.com

Library of Congress Cataloging-in-Publication Data
Saad, Michel 1927–
Mechanical discipline-specific review for the FE/EIT exam / Michel Saad, Abdie H. Tabrizi, with Michael R. Lindeburg.--2nd ed.
p. cm.
ISBN-13: 978-1-59126-065-3
ISBN-10: 1-59126-065-5
1. Engineering--United States--Examinations--Study guides. 2. Mechanical engineering--United States--Examinations--Study guides. 3. Engineering--Problems, exercises, etc. 4. Engineers--Certification--United States. I. Tabrizi, Abdie H. II. Lindeburg, Michael R. III. Title.

TA159.S23 2006
621.076--dc22

2005057891

Table of Contents

Preface and Acknowledgments

This book is one in a series intended for engineers and students who are taking a discipline-specific (DS) afternoon session of the Fundamentals of Engineering (FE) exam.

The topics covered in the DS afternoon FE exams are completely different from the topics covered in the morning session of the FE exam. Since this book only covers one discipline-specific exam, it provides exam-level problems that are like those found on the afternoon half of the FE exam for the mechanical discipline.

This book is intended to be a quick review of the material unique to the afternoon session of the mechanical engineering exam. The material presented covers the subjects most likely to be on the exam. This book is not a thorough treatment of the exam topics. Its objective is to prepare you with enough knowledge to pass. As much as practical, this book uses the notation given in the NCEES Handbook.

This book consolidates 180 practical review problems, covering all of the discipline-specific exam topics. All problems include full solutions.

The first edition of this book was developed by Michel Saad, PhD, PE, and Abdie Tabrizi, PhD, following the format, style, subject breakdown, and guidelines that I provided. Judith Vera, PE, contributed significant updates of the problems and exam sections for the second edition.

In developing this book, the NCEES Handbook and the breakdown of problem types published by NCEES were my guide for problem types and scope of coverage. However, as with most standardized tests, there is no guarantee that any specific problem type will be encountered. It is expected that minor variations in problem content will occur from exam to exam.

As with all of Professional Publications' books, the problems in this book are original and have been ethically derived. Although examinee feedback was used to determine its content, this book contains problems that are only *like* those that are on the exam. There are no actual exam problems in this book.

This book was designed to complement my *FE Review Manual*, which you will also need to prepare for the FE exam. The *FE Review Manual* is Professional Publications' most popular study guide for both the morning and afternoon general exams. It and the *Engineer-In-Training Reference Manual* have been the most popular review books for this exam for more than 25 years.

You cannot prepare adequately without your own copy of the NCEES Handbook. This document contains the data and formulas that you will need to solve both the general and the discipline-specific problems. A good way to become familiar with it is to look up the information, formulas, and data that you need while trying to work practice problems.

Exam-prep books are always works in progress. By necessity, a book will change as the exam changes. Even when the exam format doesn't change for a while, new problems and improved explanations can always be added. I encourage you to provide comments via PPI's errata reporting page, **www.ppi2pass.com/errata**. You will find all verified errata there. I appreciate all feedback.

Best of luck to you in your pursuit of licensure.

Michael Lindeburg, PE

Engineering Registration in the United States

ENGINEERING REGISTRATION

Engineering registration (also known as *engineering licensing*) in the United States is an examination process by which a state's board of engineering licensing (i.e., registration board) determines and certifies that you have achieved a minimum level of competence. This process protects the public by preventing unqualified individuals from offering engineering services.

Most engineers do not need to be registered. In particular, most engineers who work for companies that design and manufacture products are exempt from the licensing requirement. This is known as the *industrial exemption.* Nevertheless, there are many good reasons for registering. For example, you cannot offer consulting engineering design services in any state unless you are registered in that state. Even within a product-oriented corporation, however, you may find that registered engineers have more opportunities for employment and advancement.

Once you have met the registration requirements, you will be allowed to use the titles Professional Engineer (PE), Registered Engineer (RE), and Consulting Engineer (CE).

Although the registration process is similar in all 50 states, each state has its own registration law. Unless you offer consulting engineering services in more than one state, however, you will not need to register in other states.

The U.S. Registration Procedure

To become a registered engineer in the United States, you will need to pass two eight-hour written examinations. The first is the *Fundamentals of Engineering Examination*, also known as the *Engineer-In-Training Examination* and the *Intern Engineer Exam.* The initials FE, EIT, and IE are also used. This exam covers basic subjects from the mathematics, physics, chemistry, and engineering classes you took during your first four university years. In rare cases, you may be allowed to skip this first exam.

The second eight-hour exam is the *Principles and Practice of Engineering Exam.* The initials PE are also used. This exam is on topics within a specific discipline, and only covers subjects that fall within that area of specialty.

Most states have similar registration procedures. However, the details of registration qualifications, experience requirements, minimum education levels, fees, oral interviews, and exam schedules vary from state to state. For more information, contact your state's registration board (**www.ppi2pass.com/stateboards.html**).

National Council of Examiners for Engineering and Surveying

The National Council of Examiners for Engineering and Surveying (NCEES) in Clemson, South Carolina, produces, distributes, and scores the national FE and PE exams. The individual states purchase the exams from NCEES and administer them themselves. NCEES does not distribute applications to take the exams, administer the exams or appeals, or notify you of the results. These tasks are all performed by the states.

Reciprocity Among States

With minor exceptions, having a license from one state will not permit you to practice engineering in another state. You must have a professional engineering license from each state in which you work. For most engineers, this is not a problem, but for some, it is. Luckily, it is not too difficult to get a license from every state you work in once you have a license from one state.

All states use the NCEES exams. If you take and pass the FE or PE exam in one state, your certificate will be honored by all of the other states. Although there may be other special requirements imposed by a state, it will not be necessary to retake the FE and PE exams. The issuance of an engineering license based on another state's license is known as *reciprocity* or *comity.*

The simultaneous administration of identical exams in all states has led to the term *uniform examination.* However, each state is still free to choose its own minimum passing score and to add special questions and requirements to the examination process. Therefore, the use of a uniform exam has not, by itself, ensured reciprocity among states.

THE FE EXAM

Applying for the Exam

Each state charges different fees, specifies different requirements, and uses different forms to apply for the exam. Therefore, it will be necessary to request an application from the state in which you want to become registered. Generally, it is sufficient for you to phone for this application. You'll find contact information (websites, telephone numbers, email addresses, etc.) for all U.S. state and territorial boards of registration at **www.ppi2pass.com/stateboards.html**.

Keep a copy of your exam application, and send the original application by certified mail, requesting a delivery receipt. Keep your proof of mailing and delivery with your copy of the application.

Exam Dates

The national FE and PE exams are administered twice a year (usually in mid-April and late October), on the same weekends in all states. For a current exam schedule, check **www.ppi2pass.com/fefaqs.html**.

FE Exam Format

The NCEES Fundamentals of Engineering examination has the following format and characteristics.

- There are two four-hour sessions separated by a one-hour lunch.

- Examination questions are distributed in a bound examination booklet. A different exam booklet is used for each of the two sessions.

- Formulas and tables of data needed to solve questions in the exams are found in either the NCEES Handbook or in the body of the question statement itself.

- The morning session (also known as the *A.M. session*) has 120 multiple-choice questions, each with four possible answers lettered (A) through (D). Responses must be recorded with a pencil provided by NCEES on special answer sheets. No credit is given for answers recorded in ink.

- Each problem in the morning session is worth one point. The total score possible in the morning is 120 points. Guessing is valid; no points are subtracted for incorrect answers.

- There are questions on the exam from most of the undergraduate engineering degree program subjects. Questions from the same subject are all grouped together, and the subjects are labeled. The percentages of questions for each subject in the morning session are given in the following table.

Morning FE Exam Subjects

subject	percentage of questions (%)
chemistry	9
computers	7
electricity and magnetism	9
engineering economics	8
engineering probability and statistics	7
engineering mechanics (statics and dynamics)	10
ethics and business practices	7
fluid mechanics	7
material properties	7
mathematics	15
strength of materials	7
thermodynamics	7

- There are seven different versions of the afternoon session (also known as the *P.M. session*), six of which correspond to specific engineering disciplines: chemical, civil, electrical, environmental, industrial, and mechanical engineering.

  The seventh version of the afternoon exam is a general examination suitable for anyone, but in particular, for engineers whose specialties are not one of the other six disciplines. Though the subjects in the general afternoon exam are similar to the morning subjects, the questions are more complex—hence their double weighting. Questions on the afternoon exam are intended to cover concepts learned in the last two years of a four-year degree program. Unlike morning questions, these questions may deal with more than one basic concept per question.

  Each version of the afternoon session consists of 60 questions. All questions are mandatory. Questions in each subject may be grouped into related problem sets containing between two and ten questions each.

  The percentages of questions for each subject in the general afternoon session exam are given in the following table.

Afternoon FE Exam Subjects (General Exam)

subject	percentage of questions (%)
advanced engineering mathematics	10
application of engineering mechanics	13
biology	5
electricity and magnetism	12
engineering economics	10
engineering probability and statistics	9
engineering of materials	11
fluids	15
thermodynamics and heat transfer	15

Each of the discipline-specific afternoon examinations covers a substantially different body of knowledge than the morning exam. The percentages of questions for each subject in the mechanical discipline-specific afternoon session exam are as follows.

Afternoon FE Exam Subjects (Mechanical DS Exam)

subject	percentage of questions (%)
mechanical design and analysis	15
kinematics, dynamics, and vibrations	15
materials and processing	10
measurements, instrumentation, and controls	10
thermodynamics and energy conversion processes	15
fluid mechanics and fluid machinery	15
heat transfer	10
refrigeration and HVAC	10

Some afternoon questions stand alone, while others are grouped together, with a single problem statement that describes a situation followed by two or more questions about that situation. All questions are multiple choice. You must choose the best answer from among four, lettered (A) through (D).

- Each question in the afternoon is worth two points, making the total possible score 120 points.
- The scores from the morning and afternoon sessions are added together to determine your total score. No points are subtracted for guessing or incorrect answers. Both sessions are given equal weight. It is not necessary to achieve any minimum score on either the morning or afternoon sessions.
- All grading is done by computer optical sensing.

Use of SI Units on the FE Exam

Metric questions are used in all subjects, except some civil engineering and surveying subjects that typically use only customary U.S. (i.e., English) units. SI units are consistent with ANSI/IEEE standard 268 (the American Standard for Metric Practice). Non-SI metric units might still be used when common or where needed for consistency with tabulated data (e.g., use of bars in pressure measurement).

Grading and Scoring the FE Exam

The FE exam is not graded on the curve, and there is no guarantee that a certain percentage of examinees will pass. Rather, NCEES uses a modification of the Angoff procedure to determine the suggested passing score (the cutoff point or cut score).

With this method, a group of engineering professors and other experts estimate the fraction of minimally qualified engineers who will be able to answer each question correctly. The summation of the estimated fractions for all test questions becomes the passing score. Because the law in most states requires engineers to achieve a score of 70% to become licensed, you may be reported as having achieved a score of 70% if your raw score is greater than the passing score established by NCEES, regardless of the raw percentage. The actual score may be slightly more or slightly less than 110 as determined from the performance of all examinees on the equating subtest.

About 20% of the FE exam questions are repeated from previous exams—this is the *equating subtest.* Since the scores of previous examinees on the equating subtest are known, comparisons can be made between the two exams and examinee populations. These comparisons are used to adjust the passing score.

The individual states are free to adopt their own passing score, but all adopt NCEES's suggested passing score because the states believe this cutoff score can be defended if challenged.

You will receive the results within 12 weeks of taking the exam. If you pass, you will receive a letter stating that you have passed. If you fail, you will be notified that you failed and be provided with a diagnostic report.

Permitted Reference Material

Since October 1993, the FE exam has been what NCEES calls a "limited-reference" exam. This means that no books or references other than those supplied by NCEES may be used. Therefore, the FE exam is really an "NCEES-publication only" exam. NCEES provides its own Supplied-Reference Handbook for use during the examination. No books from other publishers may be used.

CALCULATORS

In most states, battery- and solar-powered, silent calculators can be used during the exam, although printers cannot be used. (Solar-powered calculators are preferable because they do not have batteries that run down.) In most states, programmable, preprogrammed, and business/finance calculators are allowed. Similarly, nomographs and specialty slide rules are permitted. To prevent unauthorized transcription and redistribution of the exam questions, calculators with communication or text-editing capabilities are banned from all NCEES exam sites. You cannot share calculators with other examinees. For a list of allowed calculators check **www.ppi2pass.com/calculators.html**.

It is essential that a calculator used for engineering examinations have the following functions.

- trigonometric functions
- inverse trigonometric functions
- hyperbolic functions
- pi
- square root and x^2
- common and natural logarithms
- y^x and e^x

For maximum speed, your calculator should also have or be programmed for the following functions.

- extracting roots of quadratic and higher-order equations
- converting between polar (phasor) and rectangular vectors
- finding standard deviations and variances
- calculating determinants of 3 × 3 matrices
- linear regression
- economic analysis and other financial functions

STRATEGIES FOR PASSING THE FE EXAM

The most successful strategy for passing the FE exam is to prepare in all of the exam subjects. Do not limit the number of subjects you study in hopes of finding enough questions in your strongest areas of knowledge to pass.

Fast recall and stamina are essential to doing well. You must be able to quickly recall solution procedures, formulas, and important data. You will not have time during the exam to derive solutions methods—you must know them instinctively. This ability must be maintained for eight hours. Be sure to gain familiarity with the NCEES Handbook by using it as your only reference for some of the problems you work during study sessions.

In order to get exposure to all exam subjects, it is imperative that you develop and adhere to a review schedule. If you are not taking a classroom review course (where the order of your preparation is determined by the lectures), prepare your own review schedule.

There are also physical demands on your body during the exam. It is very difficult to remain alert and attentive for eight hours or more. Unfortunately, the more time you study, the less time you have to maintain your physical condition. Thus, most examinees arrive at the exam site in peak mental condition but in deteriorated physical condition. While preparing for the FE exam is not the only good reason for embarking on a physical conditioning program, it can serve as a good incentive to get in shape.

It will be helpful to make a few simple decisions prior to starting your review. You should be aware of the different options available to you. For example, you should decide early on to

- use SI units in your preparation
- perform electrical calculations with effective (rms) or maximum values
- take calculations out to a maximum of four significant digits
- prepare in all exam subjects, not just your specialty areas

At the beginning of your review program, you should locate a spare calculator. It is not necessary to buy a spare if you can arrange to borrow one from a friend or the office. However, if possible, your primary and spare calculators should be identical. If your spare calculator is not identical to the primary calculator, spend some time familiarizing yourself with its functions.

A Few Days Before the Exam

There are a few things you should do a week or so before the exam date. For example, visit the exam site in order to find the building, parking areas, examination room, and rest rooms. You should also make arrangements now for child care and transportation. Since the exam does not always start or end at the designated times, make sure that your child care and transportation arrangements can tolerate a late completion.

Next in importance to your scholastic preparation is the preparation of your two examination kits. The first kit consists of a bag or box containing items to bring with you into the examination room.

[ ] letter admitting you to the exam
[ ] photographic identification
[ ] main calculator
[ ] spare calculator
[ ] extra calculator batteries
[ ] unobtrusive snacks
[ ] travel pack of tissues
[ ] headache remedy
[ ] $2.00 in change
[ ] light, comfortable sweater
[ ] loose shoes or slippers
[ ] handkerchief
[ ] cushion for your chair
[ ] small hand towel
[ ] earplugs
[ ] wristwatch with alarm
[ ] wire coat hanger
[ ] extra set of car keys

The second kit consists of the following items and should be left in a separate bag or box in your car in case you need them.

- [ ] copy of your application
- [ ] proof of delivery
- [ ] this book
- [ ] other references
- [ ] regular dictionary
- [ ] scientific dictionary
- [ ] course notes in three-ring binders
- [ ] instruction booklets for all your calculators
- [ ] light lunch
- [ ] beverages in thermos and cans
- [ ] sunglasses
- [ ] extra pair of prescription glasses
- [ ] raincoat, boots, gloves, hat, and umbrella
- [ ] street map of the exam site
- [ ] note to the parking patrol for your windshield explaining where you are, what you are doing, and why your time may have expired
- [ ] battery-powered desk lamp

The Day Before the Exam

Take the day before the exam off from work to relax. Do not cram the last night. A good prior night's sleep is the best way to start the exam. If you live far from the exam site, consider getting a hotel room in which to spend the night.

Make sure your exam kits are packed and ready to go.

The Day of the Exam

You should arrive at least 30 minutes before the exam starts. This will allow time for finding a convenient parking place, bringing your materials to the exam room, and making room and seating changes. Be prepared, though, to find that the examination room is not open or ready at the designated time.

Once the examination has started, consider the following suggestions.

- Set your wristwatch alarm for five minutes before the end of each four-hour session, and use that remaining time to guess at all of the remaining unsolved problems. Do not work up until the very end. You will be successful with about 25% of your guesses, and these points will more than make up for the few points you might earn by working during the last five minutes.
- Do not spend more than two minutes per morning question. (The average time available per problem is two minutes.) If you have not finished a question in that time, make a note of it and move on.
- Do not ask your proctors technical questions. Even if they are knowledgeable in engineering, they will not be permitted to answer your questions.
- Make a quick mental note about any problems for which you cannot find a correct response or for which you believe there are two correct answers. Errors in the exam are rare, but they do occur. Being able to point out an error later might give you the margin you need to pass. Since such problems are almost always discovered during the scoring process and discounted from the exam, it is not necessary to tell your proctor, but be sure to mark the one best answer before moving on.
- Make sure all of your responses on the answer sheet are dark and completely fill the bubbles.

Common Questions About the DS Exam

Q: Do I have to take the DS exam?

A: Most people do not have to take the DS exam and may elect the general exam option. The state boards do not care which afternoon option you choose, nor do employers. In some cases, an examinee still in an undergraduate degree program may be required by his or her university to take a specific DS exam.

Q: Do all mechanical, civil, electrical, chemical, industrial, and environmental engineers take the DS exam?

A: Originally, the concept was that examinees from the "big five" disciplines would take the DS exam, and the general exam would be for everyone else. This remains just a concept, however. A majority of engineers in all of the disciplines apparently take the general exam.

Q: When do I elect to take the DS exam?

A: You will make your decision on the afternoon of the FE exam, when the exam booklet (containing all of the DS exams) is distributed to you.

Q: Where on the application for the FE exam do I choose which DS exam I want to take?

A: You don't specify the DS option at the time of your application.

Q: After starting to work on either the DS or general exam, can I change my mind and switch options?

A: Yes. Theoretically, if you haven't spent too much time on one exam, you can change your mind and start a different one. (You might need to obtain a new answer sheet from the proctor.)

Q: After I take the DS exam, does anyone know that I took it?

A: After you take the FE exam, only NCEES and your state board will know whether you took the DS or general exam. Such information may or may not be retained by your state board.

Q: Will my DS FE certificate be recognized by other states?

A: Yes. All states recognize passing the FE exam and do not distinguish between the DS and general afternoon portions of the FE exam.

Q: Is the DS FE certificate "better" than the general FE certificate?

A: There is no difference. No one will know which option you chose. It's not stated on the certificate you receive from your state.

Q: What is the format of the DS exam?

A: The DS exam is 4 hours long. There are 60 problems, each worth 2 points. The average time per problem is 4 minutes. Each problem is multiple choice with 4 answer choices. Most problems require the application of more than one concept (i.e., formula).

Q: Is there anything special about the way the DS exam is administered?

A: In all ways, the DS and general afternoon exam are equivalent. There is no penalty for guessing. No credit is given for scratch pad work, methods, etc.

Q: Are the answer choices close or tricky?

A: Answer choices are not particularly close together in value, so the number of significant digits is not going to be an issue. Wrong answers, referred to as "distractors" by NCEES, are credible. However, the exam is not "tricky"; it does not try to mislead you.

Q: Are any problems in the afternoon session related to each other?

A: Several questions may refer to the same situation or figure. However, NCEES has tried to make all of the questions independent. If you make a mistake on one question, it shouldn't carry over to another.

Q: Is there any minimum passing score for the DS exam?

A: No. It is the total score from your morning and afternoon sessions that determines your passing, not the individual session scores. You do not have to "pass" each session individually.

Q: Is the general portion easier, harder, or the same as the DS exams?

A: Theoretically, all of the afternoon options are the same. At least, that is the intent of offering the specific options—to reduce the variability. Individual passing rates, however, may still vary 5% to 10% from exam to exam. (Professional Publications lists the most recent passing statistics for the various DS options on its website at **www.ppi2pass.com/fepassrates.html**.)

Q: Do the DS exams cover material at the undergraduate or graduate level?

A: Like the general exam, test topics come entirely from the typical undergraduate degree program. However, the emphasis is primarily on material from the third and fourth year of your program. This may put examinees who take the exam in their junior year at a disadvantage.

Q: Do you need practical work experience to take the DS exam?

A: No.

Q: Does the DS exam also draw on subjects that are in the general exam?

A: Yes. The dividing line between general and DS topics is often indistinct.

Q: Is the DS exam in customary U.S. or SI units?

A: The DS exam is nearly entirely in SI units. A few exceptions exist for some engineering subjects (surveying, hydrology, code-based design, etc.) where current common practice uses only customary U.S. units.

Q: Does the NCEES Handbook cover everything that is on the DS exam?

A: No. You may be tested on subjects that are not in the NCEES Handbook. However, NCEES has apparently adopted an unofficial policy of providing any necessary information, data, and formulas in the stem of the question. You will not be required to memorize any formulas.

Q: Is everything in the DS portion of the NCEES Handbook going to be on the exam?

A: Apparently, there is a fair amount of reference material that isn't needed for every exam. There is no way, however, to know in advance what material is needed.

Q: How long does it take to prepare for the DS exam?

A: Preparing for the DS exam is similar to preparing for a mini PE exam. Engineers typically take two to four months to complete a thorough review for the PE exam. However, examinees who are still in their degree program at a university probably aren't going to spend more than two weeks thinking about, worrying about, or preparing for the DS exam. They rely on their recent familiarity with the subject matter.

Q: If I take the DS exam and fail, do I have to take the DS exam the next time?

A: No. The examination process has no memory.

Q: Where can I get even more information about the DS exam?

A: If you have internet access, visit the Exam FAQs and the Engineering Exam Forum at Professional Publications' website (**www.ppi2pass.com/fefaqs.html** and **www.ppi2pass.com/forums.html**, respectively).

How to Use This Book

HOW EXAMINEES CAN USE THIS BOOK

This book is divided into three parts: The first part consists of 60 representative practice problems covering all of the topics in the afternoon DS exam. Sixty problems corresponds to the number of problems in the afternoon DS exam. You may time yourself by allowing approximately 4 minutes per problem when attempting to solve these problems, but that was not my intent when designing this book. Since the solution follows directly after each problem in this section, I intended for you to read through the problems, attempt to solve them on your own, become familiar with the support material in the official NCEES Handbook, and accumulate the reference materials you think you will need for additional study.

The second and third parts of this book consists of two complete sample examinations that you can use as sources of additional practice problems or as timed diagnostic tools. They also contain 60 problems, and the number of problems in each subject corresponds to the breakdown of subjects published by NCEES. Since the solutions to these parts of the book are consolidated at the end, it was my intent that you would solve these problems in a realistic mock-exam mode.

You should use the NCEES Handbook as your only reference during the mock exams.

The morning general exam and the afternoon DS exam essentially cover two different bodies of knowledge. It takes a lot of discipline to prepare for two standardized exams simultaneously. Because of that (and because of my good understanding of human nature), I suspect that you will be tempted to start preparing for your chosen DS exam only after you have become comfortable with the general subjects. That's actually quite logical, because if you run out of time, you will still have the general afternoon exam as a viable option.

If, however, you are limited in time to only two or three months of study, it will be quite difficult to do a thorough DS review if you wait until after you have finished your general review. With a limited amount of time, you really need to prepare for both exams in parallel.

HOW INSTRUCTORS CAN USE THIS BOOK

The availability of the discipline-specific FE exam has greatly complicated the lives of review course instructors and coordinators. The general consensus is that it is essentially impossible to do justice to all of the general FE exam topics and then present a credible review for each of the DS topics. Increases in course cost, expenses, course length, and instructor pools (among many other issues) all conspire to create quite a difficult situation.

One-day reviews for each DS subject are subject-overload from a reviewing examinee's standpoint. Efforts to shuffle FE students over the parallel PE review courses meet with scheduling conflicts. Another idea, that of lengthening lectures and providing more in-depth coverage of existing topics (e.g., covering transistors during the electricity lecture), is perceived as a misuse of time by a majority of the review course attendees. Is it any wonder that virtually every FE review course in the country has elected to only present reviews for the general afternoon exam?

But, while more than half of the examinees elect to take the general afternoon exam, some may actually be required to take a DS exam. This is particularly the case in some university environments where the FE exam has become useful as an "outcome assessment tool." Thus, some method of review is still needed.

Since most examinees begin reviewing approximately two to three months before the exam (which corresponds to when most review courses begin), it is impractical to wait until the end of the general review to start the DS review. The DS review must proceed in parallel with the general review.

In the absence of parallel DS lectures (something that isn't yet occurring in too many review courses), you may want to structure your review course to provide lectures only on the general subjects. Your DS review could be assigned as "independent study," using chapters and problems from this book. Thus, your DS review would consist of distributing this book with a schedule of assignments. Your instructional staff could still provide assistance on specific DS problems, and completed DS assignments could still be recorded.

The final chapter on incorporating DS subjects into review courses has yet to be written. Like the landscape architect who waits until a well-worn path appears through the plants before placing stepping stones, we need to see how review courses do it before we can give any advice.

Nomenclature

NOMENCLATURE

a	acceleration	m/s^2
a	acceleration	m/s^2
a	distance	m
a	lattice constant	m
a	speed of sound	m/s
A	area	m^2
A/F	air-fuel ratio	kg air/kg fuel
APF	atomic packing factor	–
b	base	m
c	distance to extreme fiber	m
c	specific heat	J/kg·K
C	capacitance	F
C	centroidal distance	–
C	circumference	m
C	coefficient	–
C	constant	–
C	damping factor	–
COP	coefficient of performance	–
d	distance or thickness	m
D	diameter	m
D	diffusivity	m^2/s
D	amplitude of vibration	m
DD	degree days	°C·days
DP	degree of polymerization	–
e	coefficient of restitution	–
E	electrode potential	V
E	emissive power	W/m^2
E	modulus of elasticity	Pa
E	velocity head error	Pa
f	critical speed	Hz
f	frequency	Hz
f	friction factor	–
F	black body shape factor	–
$\mathcal{F}$	Faraday constant	A·s/mol
F	force	N
FS	factor of safety	–
g	local acceleration due to gravity	m/s^2
G	shear modulus	Pa
h	convection heat transfer coefficient	$W/m^2 \cdot K$
h	enthalpy per unit mass	J/kg
h	height or head	m
H	total head	m
H	heating value	$W \cdot hr/m^3$
h_{fg}	enthalpy of vaporization per unit mass	J/kg
I	electric current	A
I	moment of inertia	m^4
$\dot{I}$	rate of process irreversibility	W
J	polar moment of inertia	m^4
k	end-restraint constant	–
k	ratio of specific heats	–
k	spring constant	N/m
k	thermal conductivity	W/m·K
KE	kinetic energy	J
K_t	stress concentration factor	–
L	length	m
m	mass	kg
$\dot{m}$	mass flow rate	kg/s
M	Mach number	–
M	moment	N·m
MW	molecular weight	kg/kmol
n	change in oxidation state	–
n	exponent	–
n	number	–
N	normal force	N
N	number of moles or atoms	–
N	rotational speed	rpm
p	pressure	Pa
P	force	N
P	power	W
PE	potential energy	–
Pr	Prandtl number	–
q	heat transfer per unit mass	J/kJ
Q	activation energy	kJ/mol
Q	heat transfer	W
Q	total heat transfer rate	J
Q	volumetric flow rate	m^3/s
$\dot{Q}$	heat transfer rate per unit volume	W/m^3
r	radius	m
r	ratio	–
R	gas constant	J/kg·K
R	reaction force	N
R	resistance	Ω
R	specific gas constant	J/kg·K
R	thermal resistance	K/W
$\overline{R}$	universal gas constant	J/mol·K
Ra	Rayleigh number	–

Re	Reynolds number	–
s	entropy per unit mass	J/kg·K
s	Laplace transformation independent parameter	–
S	allowable shear stress	Pa
S	entropy	J/K
S	sensitivity factor	–
SE	strain energy	J
SG	specific gravity	–
t	thickness	m
t	time	s
T	absolute temperature	K
T	temperature	°C
T	tension	N
T	torque	N·m
T	transfer function	–
u	internal energy per unit mass	J/kg
U	overall heat-transfer coefficient	W/m²·K
U	strain energy	J
v	varying voltage	V
v	velocity	m/s
v′	velocity after impact	m/s
V	vertical shear force	N
V	voltage	V
V	volume	m³
$\dot{V}$	volumetric flow rate	m³/s
w	work per unit mass	J/kg
W	rate of work	W
W	weight	N
W	work	J
x	mole fraction	–
x	quality	–
y	deflection or distance	m
z	elevation	m

SYMBOLS

α	angular acceleration	rad/s²
α	coefficient of thermal expansion	1/°C
β	coefficient of volumetric expansion	K^{-1}
β	magnification factor	–
γ	specific weight	N/m³
δ	deflection	m
δ	deformation	m
ε	emissivity	–
ε	strain	–
η	efficiency	–
θ	angle	°
λ	angle	°
λ	wavelength	m
μ	absolute viscosity	N·s/m²
μ	coefficient of friction	–
μ_k	coefficient of dynamic friction	–
μ_s	coefficient of static friction	–
ν	kinematic viscosity	m²/s
ν	Poisson ratio	–
ρ	density	kg/m³
σ	normal or tensile stress	Pa
σ	Stefan-Boltzmann constant	W/m²·K⁴
τ	shear stress	Pa
τ_r	resolved shear stress	Pa
υ	specific volume	m³/kg
ϕ	angle	°
ϕ	relative humidity	–
ψ	angle	°
ω	angular velocity	rad/s
ω	frequency	cycles/s
ω	specific humidity	kg/kg
ω_f	forcing frequency	radians
ω_n	natural frequency	cycles/s

SUBSCRIPTS

0	stagnation
12	from body 1 to body 2
a	axial or dry air
b	black body
br	brass
c	capacitor, centrifugal, clearance, compression, or cutoff
C	cold
cg	center of gravity
cp	center of pressure
cr	convection-radiation or critical
d	discharge
D	drag
db	dry bulb
dp	dew point
e	exit or external
es	exit for isentropic process
f	final, friction, or saturated liquid
fg	vapor-liquid phase
g	saturated vapor
H	hot or high temperature
i	initial or inside
l	lateral
L	low temperature
m	manometer fluid
n	natural or normal
NA	neutral axis
o	output or outside
p	constant pressure or pressure
pst	pseudo-static
r	radius or resolved
s	surface

sat	saturation
sh	superheated
st	steel
t	tangential, thermal, throat, or turbine
t_c	circumferential tensile
t_L	longitudinal tensile
t_{max}	maximum tensile
v	vapor
v	velocity
V	constant volume
w	wall or water
wb	wet bulb
∞	surroundings
v	volume compression

Practice Problems

MECHANICAL DESIGN AND ANALYSIS

Problem 1

A cantilever beam with a rectangular cross section is subjected to an inclined force, as shown.

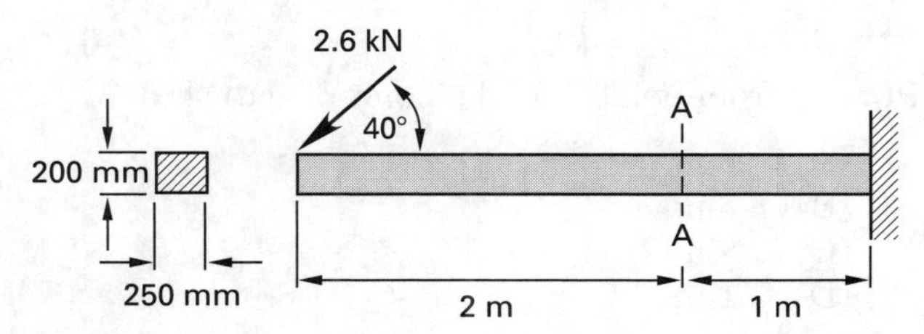

The maximum bending stress at section A-A is most nearly

(A) 0.60 MPa
(B) 1.4 MPa
(C) 2.1 MPa
(D) 2.6 MPa

Solution

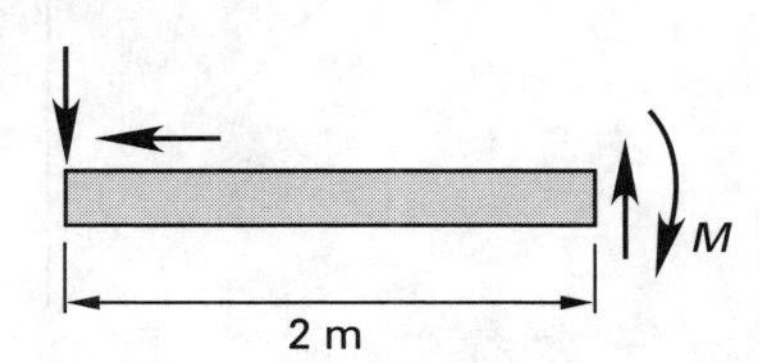

The load is eccentric.

$$\begin{aligned} M_{\text{A-A}} &= (2.6\ \text{kN})\sin 40^\circ(2\ \text{m}) \\ &\quad + (2.6\ \text{kN})\cos 40^\circ(0.1\ \text{m}) \\ &= 3.54\ \text{kN·m} \end{aligned}$$

$$\begin{aligned} \sigma_{\max} &= \frac{Mc}{I} = \frac{Mc}{\frac{1}{12}bh^3} \\ &= \frac{(3.54\times 10^3\ \text{N·m})(0.1\ \text{m})}{\left(\frac{1}{12}\right)(0.25\ \text{m})(0.2\ \text{m})^3} \\ &= 2.12\times 10^6\ \text{Pa} \quad (2.1\ \text{MPa}) \end{aligned}$$

The answer is C.

Problem 2

A slender, round column is to be designed to carry a maximum axial load of 15 kN. The column is fixed at one end and free at the other, with a total free length of 0.5 m. The modulus of elasticity and the yield strength are 200 GPa and 700 MPa, respectively. The design factor of safety is 3.5.

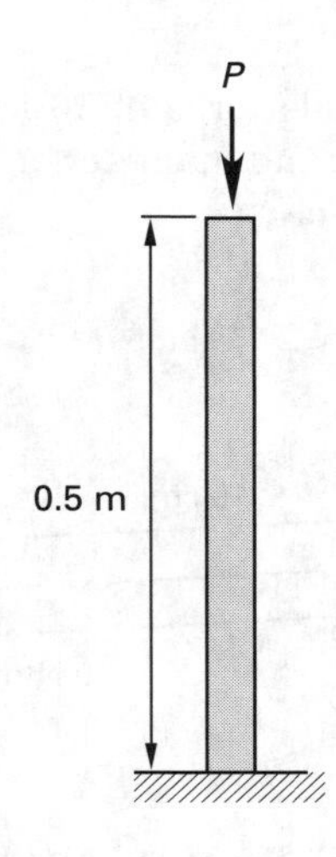

The minimum column diameter is most nearly

(A) 10 mm
(B) 20 mm
(C) 30 mm
(D) 40 mm

Solution

The design load is the critical buckling load. The end restraint constant, k, is

$$\begin{aligned} P_{\text{cr}} &= (\text{FS})P = (3.5)(15\ \text{kN}) = 52.5\ \text{kN} \\ &= \frac{\pi^2 EI}{k^2L^2} \end{aligned}$$

$$I = \left(\frac{\pi}{64}\right)D^4$$

$$\begin{aligned} D &= \left(\frac{64k^2L^2P_{\text{cr}}}{\pi^3 E}\right)^{\frac{1}{4}} \\ &= \left(\frac{(64)(2.1)^2(0.5\ \text{m})^2(52.5\ \text{kN})\left(1000\ \frac{\text{N}}{\text{kN}}\right)}{\pi^3(200\ \text{GPa})\left(10^9\ \frac{\text{Pa}}{\text{GPa}}\right)}\right)^{\frac{1}{4}} \\ &= 0.0278\ \text{m} \quad (30\ \text{mm}) \end{aligned}$$

$$\sigma = \frac{P}{A} = \frac{(52.5\text{ kN})\left(1000\ \frac{\text{N}}{\text{kN}}\right)}{\left(\frac{\pi}{4}\right)(0.0278\text{ m})^2}\left(\frac{1\text{ MPa}}{10^6\text{ Pa}}\right)$$

$$= 86\text{ MPa} < \left(\frac{1}{2}\right)(700\text{ MPa})\quad [\text{OK}]$$

The answer is C.

Problem 3

A round cantilever beam carries a vertical tip load of 2 kN, an axial tensile force of 10 kN, and a torque of 50 N·m. The length and diameter of the beam are 15 cm and 5 cm, respectively.

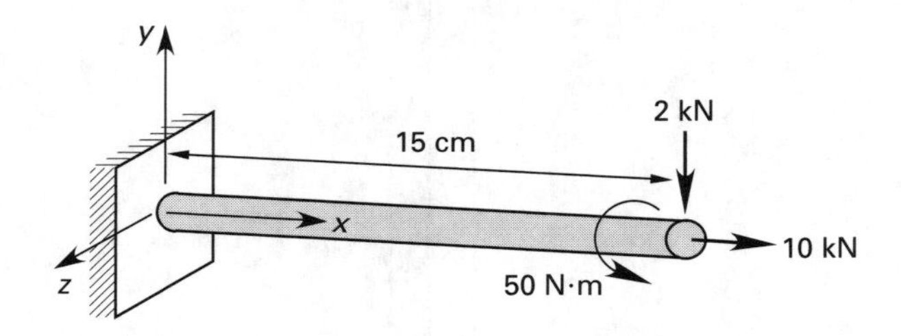

The maximum shear stress is most nearly

(A) 1.0 MPa
(B) 5.0 MPa
(C) 10 MPa
(D) 15 MPa

Solution

$$M = FL = (2\,\text{kN})(15\,\text{cm})\left(\frac{1\text{ m}}{100\text{ cm}}\right)$$
$$= 0.3\,\text{kN·m}$$

$$\tau_{\text{max}} = \left(\frac{2}{\pi D^3}\right)\sqrt{(8M + FD)^2 + (8T)^2}$$

$$= \left(\frac{2}{\pi(0.05\text{ m})^3}\right) \times \sqrt{\left(\begin{array}{c}(8)(0.3\text{ kN·m})\left(1000\ \frac{\text{N}}{\text{kN}}\right) \\ +(10\,\text{kN})\left(1000\ \frac{\text{N}}{\text{kN}}\right)(0.05\,\text{m})\end{array}\right)^2 + ((8)(50\,\text{N·m}))^2}$$

$$= 1.49 \times 10^7\text{ Pa}\quad (15\,\text{MPa})$$

The answer is D.

Problem 4

A beam of rectangular cross section is supported and loaded, as shown.

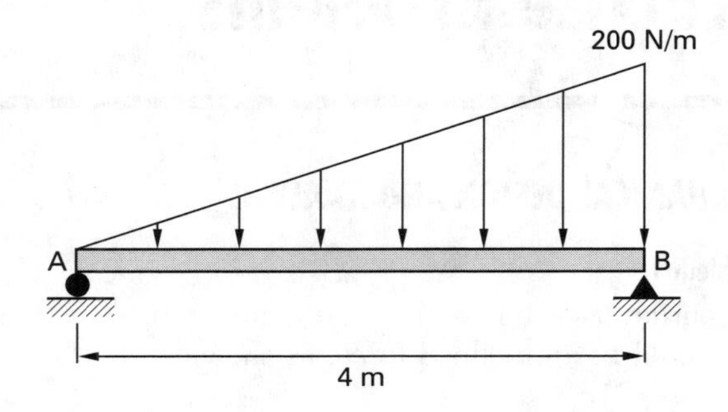

How far from point A is the point of zero shear?

(A) 1.8 m
(B) 2.3 m
(C) 2.8 m
(D) 3.1 m

Solution

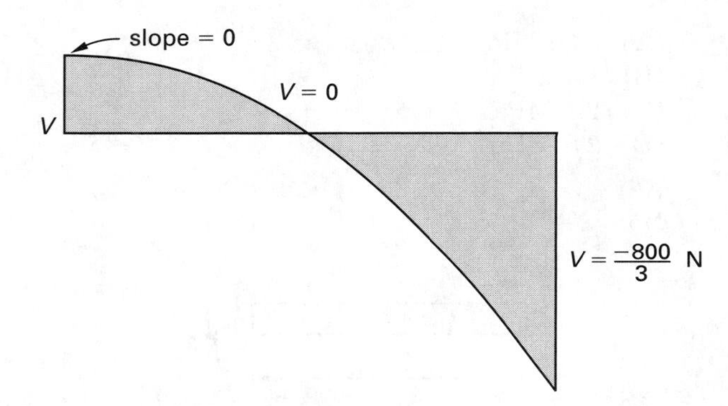

$\sum M_{\text{A}} = 0$:

$$R_{\text{B}}(4\text{ m}) - \left(\frac{1}{2}\right)\left(200\ \frac{\text{N}}{\text{m}}\right)(4\text{ m})(4\text{ m})\left(\frac{2}{3}\right) = 0$$

$$R_{\text{B}} = \frac{800}{3}\text{ N}$$

$$R_{\text{A}} = \left(\frac{1}{2}\right)\left(200\ \frac{\text{N}}{\text{m}}\right)(4\text{ m}) - \frac{800}{3}\text{ N} = \frac{400}{3}\text{ N}$$

$\sum F_y = 0$:

$$\frac{400}{3}\text{ N} - \left(\frac{1}{2}\right)\left(200\ \frac{\text{N}}{\text{m}}\right)\left(\frac{x}{4\text{ m}}\right)x = 0$$

$$x = 2.31\text{ m}\quad (2.3\text{ m})$$

The answer is B.

Problem 5

A bolt of 25 mm diameter is used to suspend a load. The bolt is supported by a timber beam, as shown. The maximum tensile stress developed in the bolt is 83 MPa. A rigid circular steel washer is used to lower the bearing stress on the beam to a maximum value of 3.4 MPa. The bolt fits snugly through the washer.

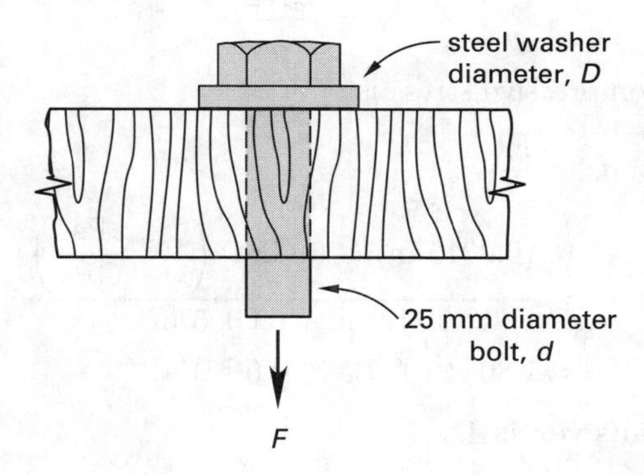

The minumum diameter for the circular washer should be most nearly

(A) 75 mm
(B) 100 mm
(C) 130 mm
(D) 150 mm

Solution

$$F = \sigma_{\text{bolt}} A_{\text{bolt}} = \sigma_{\text{bolt}} \frac{\pi}{4} d^2$$
$$= (83\ \text{MPa}) \left(10^6\ \frac{\text{Pa}}{\text{MPa}}\right) \left(\frac{\pi}{4}\right) \times \left((25\ \text{mm}) \left(\frac{1\ \text{m}}{1000\ \text{mm}}\right)\right)^2$$
$$= 40\,743\ \text{N}$$

$$A_{\text{bearing}} = \frac{F}{\sigma_{\text{bearing}}} = \frac{40\,743\ \text{N}}{(3.4\ \text{MPa}) \left(10^6\ \frac{\text{Pa}}{\text{MPa}}\right)} = 1.198 \times 10^{-2}\ \text{m}^2$$

The net contact bearing area is

$$A_{\text{net}} = \frac{\pi}{4} D^2 - \frac{\pi}{4} d^2$$

Assume the washer hole diameter is the same as the bolt diameter.

$$1.198 \times 10^{-2}\ \text{m}^2 = \frac{\pi}{4} D^2 - \frac{\pi}{4} (25\ \text{mm}) \left(\frac{1\ \text{m}}{1000\ \text{mm}}\right)$$
$$D = 0.126\ \text{m} \quad (130\ \text{mm})$$

The answer is C.

Problem 6

Three aluminum bar segments, A, B, and C, are loaded as shown. The diameters of the round segments A and B are 2.5 cm and 4 cm, respectively, and the square cross section of segment C is 5 cm × 5 cm.

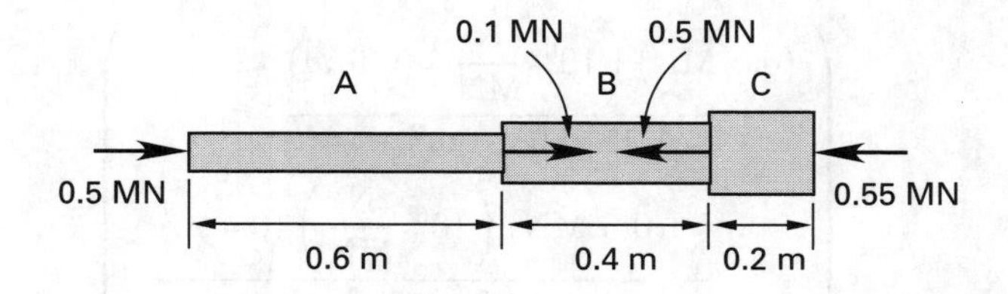

The total ideal deformation of the entire system is most nearly

(A) 0.001 m
(B) 0.005 m
(C) 0.01 m
(D) 0.02 m

Solution

The free-body diagrams are as follows.

Segment A:

Segment B:

Segment C:

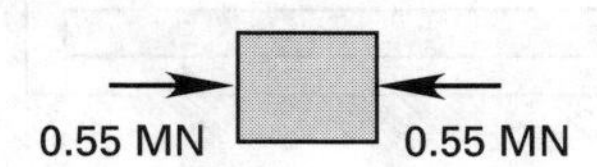

Areas:

$$A_{\text{A}} = \frac{\pi}{4} D_{\text{A}}^2 = \left(\frac{\pi}{4}\right) \left((2.5\ \text{cm}) \left(\frac{1\ \text{m}}{100\ \text{cm}}\right)\right)^2 = 4.91 \times 10^{-4}\ \text{m}^2$$

$$A_{\text{B}} = \frac{\pi}{4} D_{\text{B}}^2 = \left(\frac{\pi}{4}\right) \left((4\ \text{cm}) \left(\frac{1\ \text{m}}{100\ \text{cm}}\right)\right)^2 = 1.26 \times 10^{-3}\ \text{m}^2$$

$$A_{\text{C}} = (5\ \text{cm}) \left(\frac{1\ \text{m}}{100\ \text{cm}}\right) (5\ \text{cm}) \left(\frac{1\ \text{m}}{100\ \text{cm}}\right) = 2.5 \times 10^{-3}\ \text{m}^2$$

The total compression is $\delta = \delta_A + \delta_B + \delta_C$.

$$\delta = \sum \frac{P_i L_i}{A_i E_i}$$

For aluminum, $E = 69$ GPa.

$$\delta = \left(\begin{array}{l} \dfrac{(0.5\text{ MN})\left(10^6 \dfrac{\text{N}}{\text{MN}}\right)(0.6\text{ m})}{4.91\times10^{-4}\text{ m}^2} \\ + \dfrac{(0.6\text{ MN})\left(10^6 \dfrac{\text{N}}{\text{MN}}\right)(0.4\text{ m})}{1.26\times10^{-3}\text{ m}^2} \\ + \dfrac{(0.55\text{ MN})\left(10^6 \dfrac{\text{N}}{\text{MN}}\right)(0.2\text{ m})}{2.5\times10^{-3}\text{ m}^2} \end{array} \right) \times \left(\frac{1}{(69\text{ GPa})\left(10^9 \dfrac{\text{Pa}}{\text{GPa}}\right)} \right)$$

$$= 0.0123\text{ m} \quad (0.01\text{ m})$$

The answer is C.

Problem 7

A bar of length 1.0 m is placed horizontally, as shown. The gap between the right end of the bar and the rigid right wall is 0.5 mm. The coefficient of thermal expansion, α, and the modulus of elasticity, E, of the bar are 20×10^{-6} 1/°C and 120 GPa, respectively.

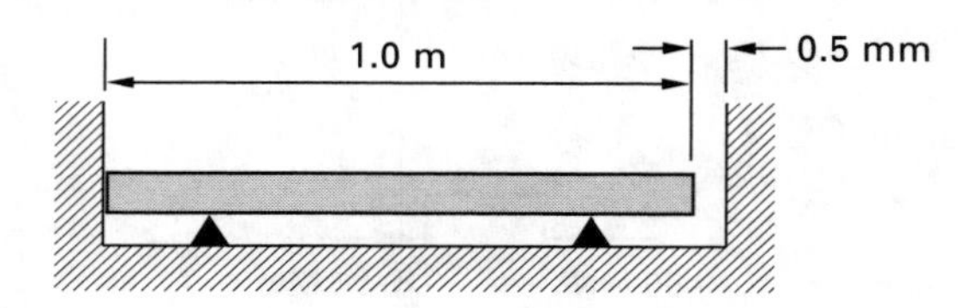

If the temperature of only the bar is raised by 100°C, the axial compressive stress produced in the bar as a result of elongation will most nearly be

(A) 50 MPa
(B) 75 MPa
(C) 120 MPa
(D) 180 MPa

Solution

The unconfined elongation due to the temperature rise is

$$\begin{aligned} \delta &= \alpha \Delta T L \\ &= \left(20 \times 10^{-6} \frac{1}{^\circ\text{C}}\right)(100^\circ\text{C})(1.0\text{ m}) \\ &= 2 \times 10^{-3}\text{ m} \quad (2\text{ mm}) \end{aligned}$$

The compression is

$$\begin{aligned} \delta_c &= \delta - 0.5\text{ mm} \\ &= 2\text{ mm} - 0.5\text{ mm} \\ &= 1.5\text{ mm} \quad (0.0015\text{ m}) \end{aligned}$$

The compressive force is

$$F_c = \frac{\delta_c A E}{L}$$

The compressive stress is

$$\begin{aligned} \sigma_c &= \frac{F_c}{A} = \frac{\delta_c E}{L} \\ &= \frac{(0.0015\text{ m})(120\text{ GPa})\left(10^9 \dfrac{\text{Pa}}{\text{GPa}}\right)}{1.0\text{ m} + 0.0005\text{ m}} \\ &= 1.80 \times 10^8\text{ Pa} \quad (180\text{ MPa}) \end{aligned}$$

The answer is D.

Problem 8

A motor delivers 500 kW to the shaft and gear assembly shown. The shaft speed is 4 Hz, and gears A and B transmit 300 kW and 200 kW to their respective mechanisms. The allowable shear stress and the maximum angle of twist of the shaft are 60 MPa and 1°, respectively. The shear modulus of the shaft is 80 GPa.

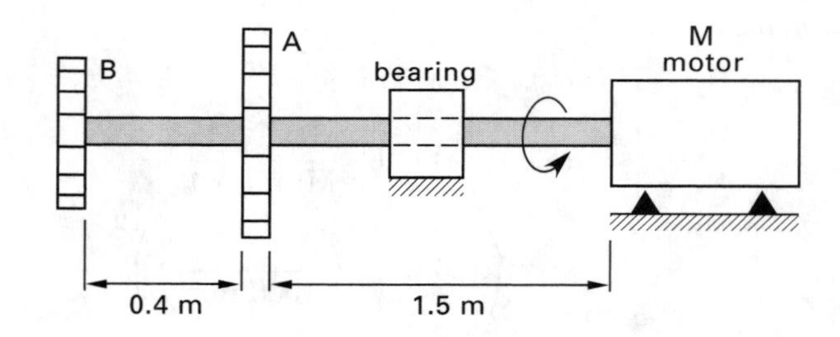

The minimum shaft diameter is most nearly

(A) 80 mm
(B) 120 mm
(C) 130 mm
(D) 150 mm

Solution

The largest torque occurs in the shorter shaft section.

$$\begin{aligned} T_1 &= \frac{P}{2\pi f} \\ &= \frac{(500\text{ kW})\left(10^3 \dfrac{\text{W}}{\text{kW}}\right)}{2\pi(4\text{ Hz})} \\ &= 1.989 \times 10^4\text{ N·m} \end{aligned}$$

$$\tau = \frac{16 T_1}{\pi D^3}$$

$$D = \sqrt[3]{\frac{16T_1}{\pi\tau}}$$
$$= \sqrt[3]{\frac{(16)(1.989 \times 10^4 \text{ N·m})}{\pi(60 \text{ MPa})\left(10^6 \frac{\text{Pa}}{\text{MPa}}\right)}}$$
$$= 0.119 \text{ m} \quad (119 \text{ mm}) \quad \text{[as limited by torque]}$$

$$T_2 = \frac{(200 \text{ kW})\left(10^3 \frac{\text{W}}{\text{kW}}\right)}{2\pi(4 \text{ Hz})}$$
$$= 7.96 \times 10^3 \text{ N·m}$$

$$J = \frac{\pi D^4}{32}$$

$$\phi = \sum_{i=1}^{2} \frac{T_i L_i}{G J_i}$$
$$= \left(\frac{32}{G\pi D^4}\right) \sum_{1}^{2} T_i L_i$$

$$(1^\circ)\left(\frac{2\pi \text{ rad}}{360^\circ}\right) = \left(\frac{32}{(80 \text{ GPa})\left(10^9 \frac{\text{Pa}}{\text{GPa}}\right)\pi D^4}\right) \times \left(\begin{array}{l}(1.989 \times 10^4 \text{ N·m})(1.5 \text{ m}) \\ + (7.96 \times 10^3 \text{ N·m})(0.4 \text{ m})\end{array}\right)$$

Solving for D,

$$D = 0.125 \text{ m} \quad (130 \text{ mm}) \quad \text{[as limited by twist]}$$

Twist is the limiting factor.

The answer is C.

Problem 9

The round cantilever beam ABC is stepped from $d_1 = 2$ cm to $d_2 = 1$ cm. The length of each section is 0.5 m, and the modulus of elasticity is 70 GPa.

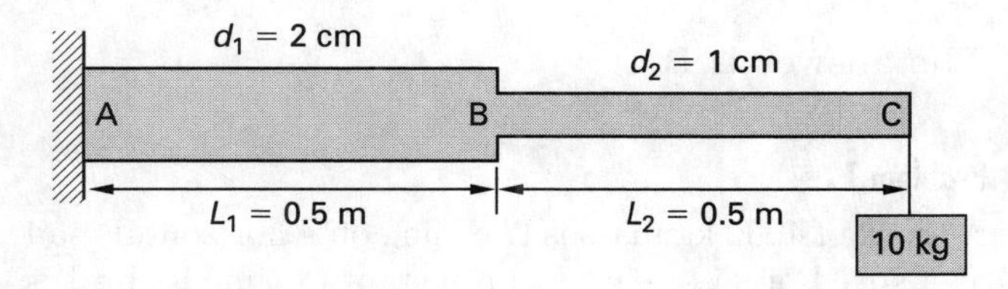

If a mass of 10 kg is supported at the free end, the deflection of the tip is most nearly

(A) 100 mm
(B) 140 mm
(C) 170 mm
(D) 200 mm

Solution

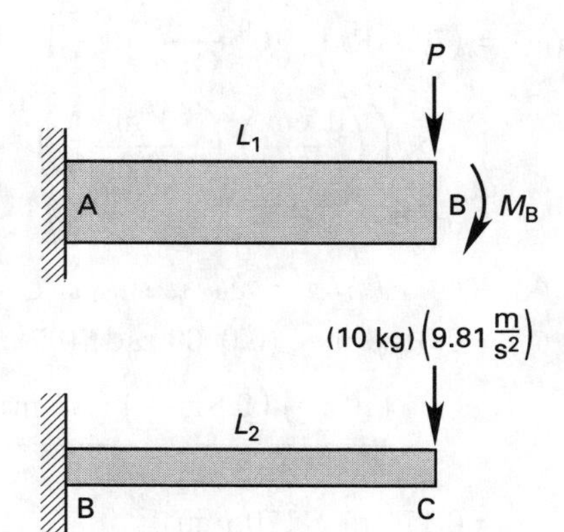

$$EI = (70 \text{ GPa})\left(10^9 \frac{\text{Pa}}{\text{GPa}}\right)\left(\frac{\pi}{4}\right) \times \left(\left(\frac{2 \text{ cm}}{2}\right)\left(\frac{1 \text{ m}}{100 \text{ cm}}\right)\right)^4$$
$$= 549.8 \text{ N·m}^2$$

The moment at B due to the mass is

$$M_B = FL_1 = mgL_1$$
$$= (10 \text{ kg})\left(9.81 \frac{\text{m}}{\text{s}^2}\right)(0.5 \text{ m})$$
$$= 49.05 \text{ N·m}$$

The deflection at B is

$$\delta_B = \delta_{\text{due to force at B}} + \delta_{\text{due to moment at B}}$$

From the beam deflection equations,

$$\delta_B = \left(\frac{FL_1^2}{6EI}\right)(3L_1 - L_1) + \frac{M_B L_1^2}{2EI}$$
$$= \frac{(10 \text{ kg})\left(9.81 \frac{\text{m}}{\text{s}^2}\right)(0.5 \text{ m})^3}{(3)(549.8 \text{ N·m}^2)} + \frac{(49.05 \text{ N·m})(0.5 \text{ m})^2}{(2)(549.8 \text{ N·m}^2)}$$
$$= 0.0186 \text{ m}$$

The angle of deflection at B is

$$\phi_B = \phi_{\text{due to force}} + \phi_{\text{due to moment}}$$
$$= \frac{FL_1^2}{2EI} + \frac{M_B L_1}{EI}$$
$$= \frac{(10 \text{ kg})\left(9.81 \frac{\text{m}}{\text{s}^2}\right)(0.5 \text{ m})^2}{(2)(549.8 \text{ N·m}^2)} + \frac{(49.05 \text{ N·m})(0.5 \text{ m})}{549.8 \text{ N·m}}$$
$$= 0.0669 \text{ rad}$$

$$\delta_{\text{due to force at C}} = \frac{FL_2^3}{3EI_{BC}}$$

$$EI_{BC} = (70\ \text{GPa})\left(10^9\ \frac{\text{Pa}}{\text{GPa}}\right)\left(\frac{\pi}{4}\right) \times \left(\left(\frac{1\ \text{cm}}{2}\right)\left(\frac{1\ \text{m}}{100\ \text{cm}}\right)\right)^4 = 34.36\ \text{N·m}^2$$

$$\delta_C = \delta_B + \theta_B L_2 + \delta_{\text{due to force at C}} = 0.0186\ \text{m} + (0.0669\ \text{rad})(0.5\ \text{m}) + \frac{(10\ \text{kg})\left(9.81\ \frac{\text{m}}{\text{s}^2}\right)(0.5\ \text{m})^3}{(3)(34.36\ \text{N·m}^2)} = 0.171\ \text{m}\quad (170\ \text{mm})$$

The answer is C.

KINEMATICS, DYNAMICS, AND VIBRATIONS

Problem 10

A 10 g ball is released vertically from a height of 10 m. The ball strikes a horizontal surface and bounces back. The coefficient of restitution between the surface and the ball is 0.75. The height that the ball will reach after bouncing is most nearly

(A) 3.5 m
(B) 5.6 m
(C) 8.5 m
(D) 11 m

Solution

The velocity of the ball just prior to impact can be found from the conservation of energy principle.

$$\text{KE} = PE$$
$$\tfrac{1}{2}m\text{v}_0^2 = mgh_0$$
$$\text{v}_0 = \sqrt{2gh_0}$$

The velocity of the ball just after impact is found from

$$e = \frac{\text{v}_r}{\text{v}_0} = \frac{\text{v}_r}{\sqrt{2gh_0}}$$

Use the conservation of energy principle to find the rebound height.

$$\tfrac{1}{2}m\text{v}_r^2 = mgh_r$$
$$h_r = \frac{\text{v}_r^2}{2g} = e^2 h_0 = (0.75)^2(10\ \text{m}) = 5.625\ \text{m}\quad (5.6\ \text{m})$$

The answer is B.

Problem 11

A 5 g mass is to be placed on a 50 cm diameter horizontal table that is rotating at 50 rpm. The mass must not slide away from its position. The coefficient of friction between the mass and the table is 0.2. What is the maximum distance that the mass can be placed from the axis of rotation?

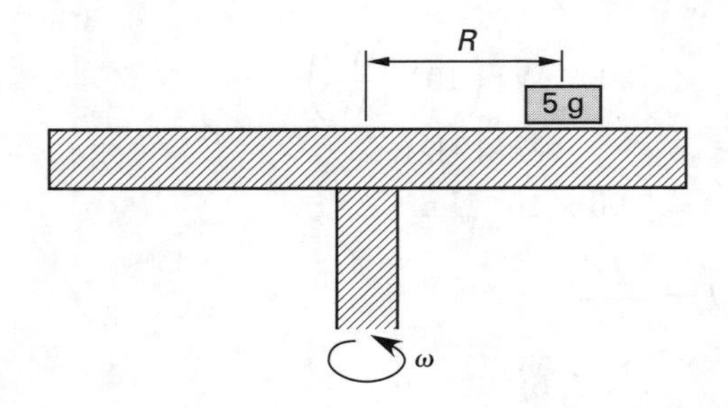

(A) 2 cm
(B) 7 cm
(C) 10 cm
(D) 20 cm

Solution

Balance the centrifugal (radial) and frictional forces.

$$\sum F = 0$$
$$F_c - F_f = 0 \qquad \text{Eq. 1}$$
$$F_c = ma_r = m\omega^2 R \qquad \text{Eq. 2}$$
$$F_f = \mu N = \mu W = \mu mg \qquad \text{Eq. 3}$$

Substitute Eq. 2 and Eq. 3 into Eq. 1 and solve for R.

$$R = \frac{\mu g}{\omega^2} = \frac{(0.2)\left(9.81\ \frac{\text{m}}{\text{s}^2}\right)}{\left(\left(50\ \frac{\text{rev}}{\text{min}}\right)\left(\frac{1\ \text{min}}{60\ \text{s}}\right)\left(2\pi\ \frac{\text{rad}}{\text{rev}}\right)\right)^2} = 0.0716\ \text{m}\quad (7\ \text{cm})$$

The answer is B.

Problem 12

A truck of 4000 kg mass is traveling on a horizontal road at a speed of 95 km/h. At an instant of time its brakes are applied, locking the wheels. The dynamic coefficient of friction between the wheels and the road is 0.42. The stopping distance of the truck is most nearly

(A) 55 m
(B) 70 m
(C) 85 m
(D) 100 m

Solution

$$\sum F = ma$$

$$-F_f = m\frac{dv}{dt} \quad \textit{Eq. 1}$$

$$F_f = \mu N = \mu mg \quad \textit{Eq. 2}$$

$$\frac{dv}{dt} = \frac{dv}{dx}\frac{dx}{dt} = v\frac{dv}{dx} \quad \textit{Eq. 3}$$

Substituting Eq. 2 and Eq. 3 into Eq. 1,

$$-\mu mg = mv\frac{dv}{dx}$$

$$-\mu g\,dx = v\,dv \quad \textit{Eq. 4}$$

Integrating Eq. 4,

$$-\mu g(x_2 - x_1) = \tfrac{1}{2}(v_2^2 - v_1^2)$$

$$x_2 = x_1 - \frac{v_2^2 - v_1^2}{2\mu g}$$

$$= 0 - \frac{0 - \left(\left(95\ \frac{\text{km}}{\text{h}}\right)\left(1000\ \frac{\text{m}}{\text{km}}\right)\left(\frac{1\ \text{h}}{3600\ \text{s}}\right)\right)^2}{(2)(0.42)\left(9.81\ \frac{\text{m}}{\text{s}^2}\right)}$$

$$= 84.5\ \text{m} \quad (85\ \text{m})$$

The answer is C.

Problem 13

A translating and rotating ring of mass 1 kg, angular speed of 500 rpm, and translational speed of 1 m/s is placed on a horizontal surface. The coefficient of friction between the ring and the surface is 0.35.

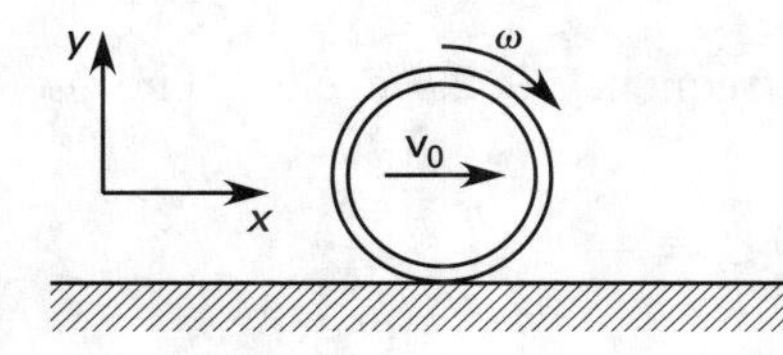

For an outside radius of 3 cm, the time at which skidding stops and rolling begins is most nearly

(A) 0.01 s
(B) 0.1 s
(C) 1 s
(D) 10 s

Solution

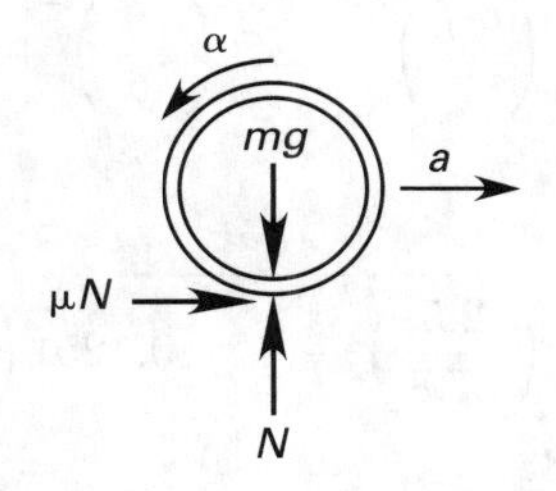

$$\sum F_x = ma$$

$$\mu N = ma \quad \textit{Eq. 1}$$

$$\sum F_y = 0$$

$$N = mg \quad \textit{Eq. 2}$$

$$\sum M = I\alpha$$

$$\mu NR = 3mR^2\alpha \quad \textit{Eq. 3}$$

From Eqs. 1 and 2, $a = \mu g$. From Eq. 3, $\alpha = \mu g/3R$.

$$v = v_0 + \mu gt \quad \textit{Eq. 4}$$

$$\alpha = \frac{d\omega}{dt}$$

$$\omega = \omega_0 - \left(\frac{\mu g}{3R}\right)t \quad \textit{Eq. 5}$$

When skidding stops, $v = R\omega$. Multiply Eq. 5 by R and compare with Eq. 4.

$$t = \frac{R\omega_0 - v_0}{\mu g\left(1 + \frac{1}{3}\right)}$$

$$= \frac{\left((3\ \text{cm})\left(\frac{1\ \text{m}}{100\ \text{cm}}\right)\right) \times \left(\left(500\ \frac{\text{rev}}{\text{min}}\right)\left(2\pi\ \frac{\text{rad}}{\text{rev}}\right)\left(\frac{1\ \text{min}}{60\ \text{s}}\right)\right) - 1\ \frac{\text{m}}{\text{s}}}{(0.35)\left(9.81\ \frac{\text{m}}{\text{s}^2}\right)\left(\frac{4}{3}\right)}$$

$$= 0.125\ \text{s} \quad (0.1\ \text{s})$$

The answer is B.

Problem 14

A stationary uniform rod of length 1 m is struck at its tip by a 3 kg rigid ball moving horizontally with velocity of 8 m/s as shown. The mass of the rod is 7 kg, and the coefficient of restitution between the rod and the ball is 0.75.

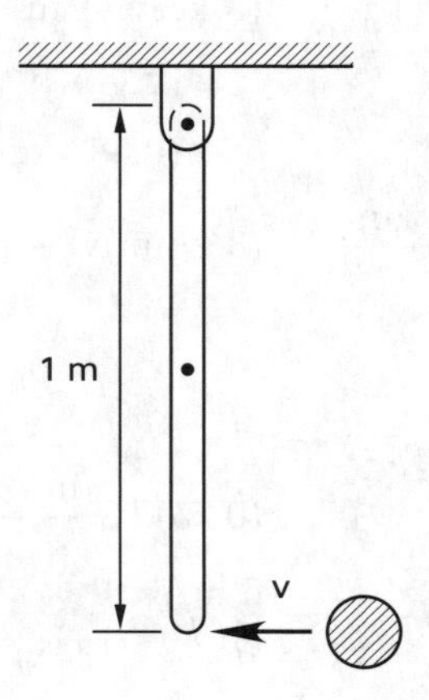

The velocity of the ball after impact is most nearly

(A) 1.9 m/s
(B) 3.5 m/s
(C) 6.0 m/s
(D) 10 m/s

Solution

Let the subscript R represent the rod.

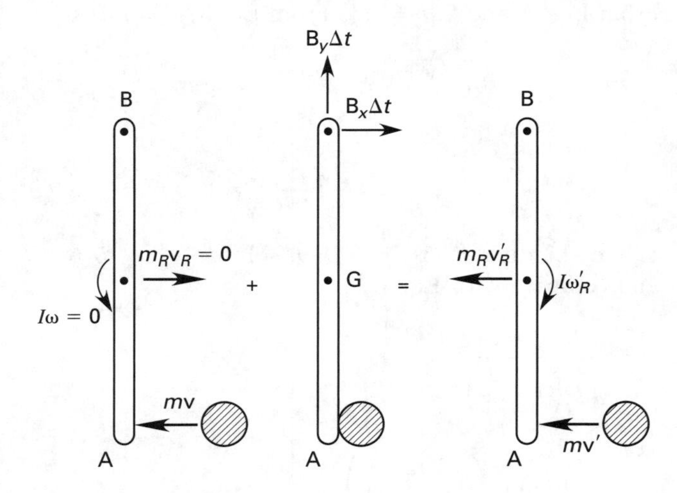

$\sum M_B$:

$$mv(1\text{ m}) = mv'(1\text{ m}) + m_R v_R'(0.5\text{ m}) + I\omega_R' \qquad \textit{Eq. 1}$$

$$v_R' = R\omega_R' = (0.5\text{ m})\omega_R'$$

$$I = \tfrac{1}{12}m_R L^2 = \left(\frac{1}{12}\right)(7\text{ kg})(1\text{ m})^2 = 0.583\text{ kg}\cdot\text{m}^2$$

Substitute values into Eq. 1.

$$(3\text{ kg})\left(8\ \frac{\text{m}}{\text{s}}\right)(1\text{ m}) = (3\text{ kg})v'(1\text{ m}) + (7\text{ kg})(0.5\text{ m})\omega_R'(0.5\text{ m}) + (0.583\text{ kg}\cdot\text{m}^2)\omega_R'$$

$$24\ \frac{\text{kg}\cdot\text{m}^2}{\text{s}} = (3\text{ kg}\cdot\text{m})v' + (2.33\text{ kg}\cdot\text{m}^2)\omega_R' \qquad \textit{Eq. 2}$$

$$v_A' - v' = e(v - v_A) = (0.75)\left(8\ \frac{\text{m}}{\text{s}} - 0\right) = 6\text{ m/s} \qquad \textit{Eq. 3}$$

$$v_A' = R\omega' = (1.0\text{ m})\omega_R' \qquad \textit{Eq. 4}$$

Solving Eqs. 2, 3, and 4 simultaneously,

$$v' = 1.88\text{ m/s} \quad (1.9\text{ m/s})$$

The answer is A.

Problem 15

A 350 kg mass, constrained to move only vertically, is supported by two springs, each having a spring constant of 250 kN/m. A periodic force with a maximum value of 100 N is applied to the mass with a frequency of 2.5 cycles. Given a damping factor of 0.125, the amplitude of the vibration is

(A) 0.198 mm
(B) 0.199 mm
(C) 0.201 mm
(D) 0.404 mm

Solution

Calculate the total spring constant.

$$k = (2)\left(250\ \frac{\text{kN}}{\text{m}}\right) = 500\text{ kN/m}$$

The natural frequency is

$$\omega = \sqrt{\frac{k}{m}} = \sqrt{\frac{\left(500\ \frac{\text{kN}}{\text{m}}\right)\left(1000\ \frac{\text{N}}{\text{kN}}\right)}{350\text{ kg}}} = 37.8\text{ rad/s}$$

The forcing frequency is 2.5 cycles. The pseudo-static deflection is

$$\delta_{\text{pst}} = \frac{F_o}{k} = \frac{100\text{ N}}{\left(500\ \frac{\text{kN}}{\text{m}}\right)\left(1000\ \frac{\text{N}}{\text{kN}}\right)} = 2\times 10^{-4}\text{ m}$$

The magnification factor is

$$\beta = \left|\frac{1}{1-\left(\frac{\omega f}{\omega}\right)^2 + 2C\left(\frac{\omega f}{\omega}\right)^2}\right| = \left|\frac{1}{1-\left(\frac{2.5}{37.8}\right)^2 + (2)(0.125)\left(\frac{2.5}{37.8}\right)^2}\right| = 1.0033$$

The amplitude of vibration is

$$\begin{aligned} D &= \beta\delta_{\text{pst}} \\ &= (1.0033)\left(2\times10^{-4}\ \text{m}\right) \\ &= \left(2.0066\times10^{-4}\ \text{m}\right)\left(1000\ \frac{\text{mm}}{\text{m}}\right) \\ &= 0.201\ \text{mm} \end{aligned}$$

The answer is C.

Problem 16

A 5 kg pendulum is swung on a 7 m long massless cord from rest at 5° from center. The time required for the pendulum to return to rest is most nearly

(A) 0.2 s
(B) 4 s
(C) 5 s
(D) 10 s

Solution

$$\begin{aligned} f &= \left(\frac{1}{2\pi}\right)\sqrt{\frac{g}{L}} = \left(\frac{1}{2\pi}\right)\sqrt{\frac{9.81\ \frac{\text{m}}{\text{s}^2}}{7\ \text{m}}} \\ &= 0.1885\ \text{s}^{-1} \\ t &= \frac{1}{f} = \frac{1}{0.1885\ \text{s}^{-1}} \\ &= 5.3\ \text{s}\quad (5\ \text{s}) \end{aligned}$$

The answer is C.

Problem 17

A garden hose is held 1.2 m off the ground at an angle of 30° from the horizontal. If the velocity of the water is 2 m/s, the horizontal distance from the nozzle head to the point where the stream hits the ground is most nearly

(A) 0.50 m
(B) 1.1 m
(C) 1.5 m
(D) 2.0 m

Solution

The y-component of velocity at time zero is

$$\begin{aligned} \text{v}_{yo} &= \left(2\ \frac{\text{m}}{\text{s}}\right)\sin 30^\circ \\ &= 1\ \text{m/s} \end{aligned}$$

The acceleration due to gravity is

$$a = -9.81\ \text{m/s}^2$$

The y-component of velocity at time t is

$$\begin{aligned} \text{v}_y &= \text{v}_{yo} + at \qquad \textit{Eq. 1} \\ &= 1\ \frac{\text{m}}{\text{s}} - \left(9.81\ \frac{\text{m}}{\text{s}^2}\right)t \end{aligned}$$

The y-component of distance at time t is

$$\begin{aligned} y &= \text{v}_{yo}t + \tfrac{1}{2}at^2 \qquad \textit{Eq. 2} \\ &= \left(1\ \frac{\text{m}}{\text{s}}\right)t - 4.9t^2 \end{aligned}$$

The y-component of velocity squared at a distance y is

$$\begin{aligned} \text{v}_y^2 &= \text{v}_{yo}^2 + 2ay \qquad \textit{Eq. 3} \\ &= 1 - 19.62y \end{aligned}$$

The x-component of velocity at time zero is

$$\begin{aligned} \text{v}_{xo} &= \left(2\ \frac{\text{m}}{\text{s}}\right)\cos 30^\circ \qquad \textit{Eq. 4} \\ &= 1.73\ \text{m/s} \end{aligned}$$

The x-component of distance is

$$\begin{aligned} x &= \text{v}_{xo}t \qquad \textit{Eq. 5} \\ &= \left(1.73\ \frac{\text{m}}{\text{s}}\right)t \end{aligned}$$

When the stream hits the ground, the y-distance is -1.2 m.

Plugging this into Eq. 2 gives

$$-1.2 = \left(1\ \frac{\text{m}}{\text{s}}\right)t + \left(\frac{1}{2}\right)\left(-9.81\ \frac{\text{m}}{\text{s}^2}\right)t^2$$

$$t^2 - 0.204t - 0.244 = 0$$

Using the quadratic formula

$$\begin{aligned} t &= \frac{-b \pm \sqrt{b^2 - 4ac}}{2a} \\ &= \frac{0.204 \pm \sqrt{(0.204)^2 - (4)(-0.244)}}{2} \\ &= 0.61\ \text{s} \end{aligned}$$

Plugging this into Eq. 5 gives

$$\begin{aligned} x &= \left(1.73\ \frac{\text{m}}{\text{s}}\right)(0.6\ \text{s}) \\ &= 1.06\ \text{m}\quad (1.1\ \text{m}) \end{aligned}$$

The answer is B.

Problem 18

A homogeneous disk of 5 cm radius and 5 kg mass rotates on an axle AB of length 0.25 m and rotates about a fixed point A. The disk is constrained to roll on a horizontal floor.

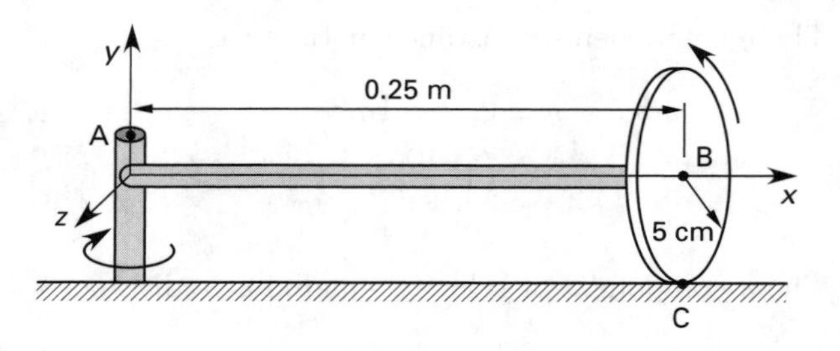

Given that the disk rotates counterclockwise at a rate of 50 rad/s about axle AB, the angular velocity of the disk is

(A) 50 rad/s in the x direction and -10 rad/s in the z direction
(B) 50 rad/s in the x direction and $+10$ rad/s in the y direction
(C) 50 rad/s in the x direction and -10 rad/s in the y direction
(D) 50 rad/s in the x direction and $+10$ rad/s in the z direction

Solution

As the disk rotates about the axle AB at a rate of $\omega_1 = 50$ rad/s it also rotates with the axle about the y-axis at a rate of ω_2 clockwise. The total angular velocity is therefore,

$$\omega = \omega_{1x} - \omega_{2z} \qquad \textit{Eq. 1}$$

To determine ω_z express the velocity of the point where the disk meets the floor, point C, equal to zero.

$$\begin{aligned} \mathrm{v}_{\text{floor}} &= \omega R_C = 0 \\ &= (\omega_{1x} - \omega_{2z})(L_x - r_y) = 0 \\ &= (L\omega_2 - r\omega_1)_z = 0 \\ \omega_2 &= \frac{r\omega_1}{L} \\ &= \frac{(5\text{ cm})\left(\frac{1\text{ m}}{100\text{ cm}}\right)\left(100\ \frac{\text{rad}}{\text{s}}\right)}{0.25\text{ m}} \\ &= 10\text{ rad/s in the } z \text{ direction} \end{aligned}$$

Substituting ω_1 into Eq. 1 gives $\omega = 50$ rad/s in the x direction -10 rad/s in the y direction.

The answer is C.

MATERIALS AND PROCESSING

Problem 19

Aluminum has a face-centered cubic (FCC) unit cell structure with a lattice constant of 0.405 nm. The density of aluminum is most nearly

(A) 0.70 g/cm^3
(B) 1.3 g/cm^3
(C) 2.7 g/cm^3
(D) 5.4 g/cm^3

Solution

$$\rho = \frac{m}{V} = \frac{(\text{no. of atoms per cell})(\text{MW})}{\text{volume of cell}}$$

$$= \frac{\left(4\ \frac{\text{atoms}}{\text{cell}}\right)\left(26.981\ \frac{\text{kg}}{\text{kmol}}\right)\left(\frac{1\text{ kmol}}{1000\text{ mol}}\right)}{\left(6.022\times10^{23}\ \frac{\text{atoms}}{\text{mol}}\right)\left((0.405\text{ nm})\left(\frac{1\text{ m}}{10^9\text{ nm}}\right)\right)^3}$$

$$= 2698\text{ kg/m}^3 \quad (2.7\text{ g/cm}^3)$$

The answer is C.

Problem 20

The diffusivity of nickel (Ni) atoms in a solid FCC iron lattice is 1.0×10^{-13} m^2/s at 1300°C and 1.0×10^{-16} m^2/s at 1000°C. The average activation energy for the diffusion of Ni atoms in the FCC iron lattice is most nearly

(A) 280 kJ/mol
(B) 380 kJ/mol
(C) 580 kJ/mol
(D) 880 kJ/mol

Solution

$$D = D_0 e^{-\frac{Q}{RT}}$$

$$\frac{D_1}{D_2} = \frac{e^{-\frac{Q}{RT_1}}}{e^{-\frac{Q}{RT_2}}} = e^{\left(\frac{Q}{R}\right)\left(\frac{-1}{T_1}+\frac{1}{T_2}\right)}$$

$$T_1 = 1300°\text{C} + 273° = 1573\text{K}$$
$$T_2 = 1000°\text{C} + 273° = 1273\text{K}$$

$$Q = \left(\frac{\overline{R}}{\frac{-1}{T_1}+\frac{1}{T_2}}\right)\ln\frac{D_1}{D_2}$$

$$= \left(\frac{8.314\ \frac{\text{J}}{\text{mol}\cdot\text{K}}}{\frac{-1}{1573\text{K}}+\frac{1}{1273\text{K}}}\right)\ln\frac{1.0\times10^{-13}\ \frac{\text{m}^2}{\text{s}}}{1.0\times10^{-16}\ \frac{\text{m}^2}{\text{s}}}$$

$$= 3.83\times10^5\text{ J/mol} \quad (380\text{ kJ/mol})$$

The answer is B.

Problem 21

454 g of solder, made of 90% lead (Pb) and 10% tin (Sn), are to be completely liquified at 200°C by the addition of more tin.

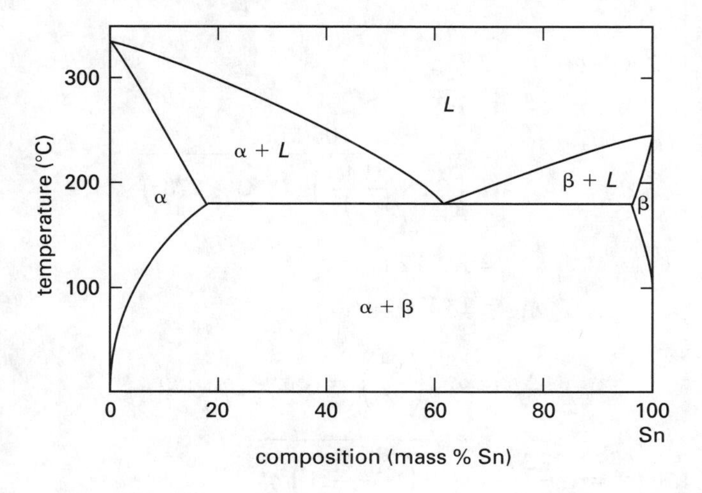

The minimum amount of tin that must be added per 100 g of solder is most nearly

(A) 114 g
(B) 125 g
(C) 140 g
(D) 165 g

Solution

Initially,

$$\begin{aligned} m_{\text{Sn}} &= (0.1)(454 \text{ g}) \\ &= 45.4 \text{ g} \end{aligned}$$

Let x equal the amount of Sn added.

From the illustration, at 200°C solder liquifies at approximately 58% Sn.

$$0.58 = \frac{m_{\text{Sn}}}{m_{\text{solder}}} = \frac{45.4 \text{ g} + x}{454 \text{ g} + x}$$

$$x = 518.9 \text{ g of Sn per 454 g solder}$$

$$\begin{aligned} m_{\text{Sn}} \text{ per 100 g solder} &= \left(\frac{518.9 \text{ g}}{454 \text{ g}}\right)(100 \text{ g}) \\ &= 114.3 \text{ g} \quad (114 \text{ g}) \end{aligned}$$

The answer is A.

Problem 22

The purpose of annealing ANSI 4160 carbon steel is to

(A) soften
(B) harden
(C) harden without cracking
(D) toughen

Solution

The purposes of annealing are to relieve internal stresses, improve the grain size, and soften the material to improve its machineability.

The answer is A.

Problem 23

A 100 cm × 100 cm steel plate is under 30 MPa stress. Assume the modulus of elasticity equals 205 GPa. The new dimension of the plate is most nearly

(A) 100 mm × 100 mm
(B) 1000 mm × 1000 mm
(C) 1000.1 × 999.1 mm
(D) 1000.1 mm × 1000 mm

Solution

Assume a Poisson's ratio, ν, of 0.29 for steel.

Poisson's ratio is defined as

$$\nu = -\frac{\varepsilon_{\text{lateral}}}{\varepsilon_{\text{axial}}}$$

Normal stress is defined as

$$\begin{aligned} \sigma &= E\varepsilon \\ \varepsilon &= \frac{\sigma}{E} \\ &= \frac{30 \text{ MPa}}{(205 \text{ GPa})\left(1000 \ \frac{\text{MPa}}{\text{GPa}}\right)} \\ \varepsilon_{\text{axial}} &= 0.0001\nu \end{aligned}$$

From Poisson's ratio,

$$\begin{aligned} \varepsilon_{\text{lateral}} &= -\nu\varepsilon_{\text{axial}} \\ &= (-0.29)\left(0.00012 \ \frac{\text{m}}{\text{m}}\right) \\ &= -0.000035 \\ &\approx 0 \end{aligned}$$

The plate in millimeters is

$$(100 \text{ cm})\left(10 \ \frac{\text{mm}}{\text{cm}}\right) = 1000 \text{ mm}$$

The axial displacement is

$$(0.0001 \text{ m})\left(1000 \ \frac{\text{mm}}{\text{m}}\right) = 0.1 \text{ mm}$$

Since the axial displacement is 0.1 mm and the lateral displacement is zero, adding these to the plate dimension gives an answer of

$$1000.1 \text{ mm} \times 1000 \text{ mm}$$

The answer is D.

Problem 24

For a zinc electrode in a solution containing 2 g of Zn^{2+} ions per liter, the electrode potential (with respect to hydrogen) is

(A) 0.67 V
(B) 0.72 V
(C) 0.76 V
(D) 0.81 V

Solution

$$\mathrm{Zn} \rightarrow \mathrm{Zn}^{2+} + 2\mathrm{e}^-$$

The atomic weight of zinc is 65.4 g.

$$E^o = 0.763 \text{ V}$$
$$E = E^o + \left(\frac{0.0257 \text{ V}}{n}\right) \ln C$$
$$C = \frac{2 \ \frac{\text{g}}{\text{L}}}{65.4 \ \frac{\text{g}}{\text{mol}}} = 0.0306 \ \frac{\text{mol}}{\text{l}}$$
$$E = 0.763 + \left(\frac{0.0257 \text{ V}}{2}\right) \ln \left(0.0306 \ \frac{\text{mol}}{\text{L}}\right) = 0.7182 \text{ V (0.72 V)}$$

The answer is B.

MEASUREMENTS, INSTRUMENTATION, AND CONTROLS

Problem 25

A pitot tube is used to measure the velocity of an air stream. The air is at a temperature of 18°C and a pressure of 105 kPa.

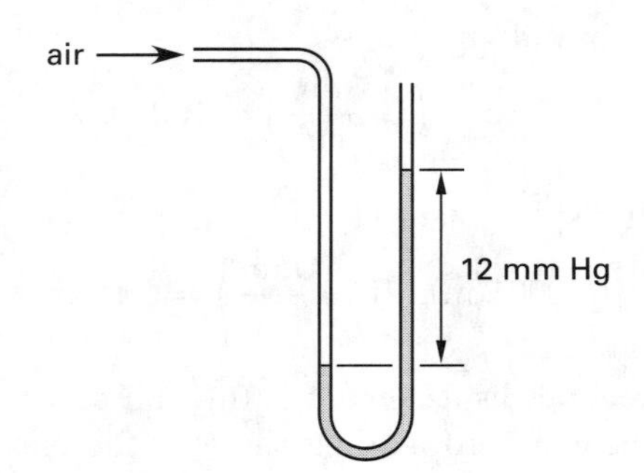

If the differential height in a mercury manometer is 12 mm and the air is incompressible, the air velocity is most nearly

(A) 38 m/s
(B) 43 m/s
(C) 48 m/s
(D) 50 m/s

Solution

$$\rho_{\text{air}} = \frac{p}{RT} = \frac{105 \text{ kPa}}{\left(0.287 \frac{\text{kJ}}{\text{kg·K}}\right)(18°\text{C} + 273°)} = 1.2572 \text{ kg/m}^3$$
$$\rho_{\text{Hg}} = 13\,550 \text{ kg/m}^3$$
$$\Delta p = \rho_{\text{air}}\left(\frac{\text{v}^2}{2}\right) = (\rho_{\text{Hg}} - \rho_{\text{air}})gh$$
$$\text{v} = \sqrt{\frac{2\Delta p}{\rho_{\text{air}}}} = \sqrt{2\left(\frac{\rho_{\text{Hg}}}{\rho_{\text{air}}} - 1\right) gh}$$
$$= \sqrt{(2)\left(\frac{13\,550 \ \frac{\text{kg}}{\text{m}^3}}{1.2572 \ \frac{\text{kg}}{\text{m}^3}} - 1\right)\left(9.81 \ \frac{\text{m}}{\text{s}^2}\right)(0.012 \text{ m})}$$
$$= 50.37 \text{ m/s} \quad (50 \text{ m/s})$$

The answer is D.

Problem 26

An oil with a specific gravity of 0.92 is in a tank pressurized to 25 kPa (gage). The oil flows through a sharp-edged orifice with a diameter of 1 cm, 3 m below the oil surface.

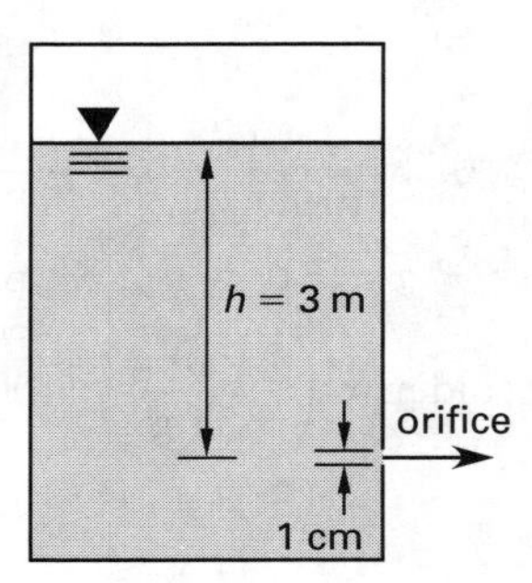

If the coefficient of discharge is 0.72, the discharge rate is most nearly

(A) 4×10^{-4} m³/s
(B) 6×10^{-4} m³/s
(C) 8×10^{-4} m³/s
(D) 10×10^{-4} m³/s

Solution

$$Q = C_d A \sqrt{2g\left(h + \frac{p}{\rho g}\right)}$$

$$= (0.72)\left(\frac{\pi}{4}\right)\left((1\ \text{cm})\left(\frac{1\ \text{m}}{100\ \text{cm}}\right)\right)^2 \times \sqrt{(2)\left(9.81\ \frac{\text{m}}{\text{s}^2}\right) \times \left(3\ \text{m} + \frac{(25\ \text{kPa})\left(1000\ \frac{\text{Pa}}{\text{kPa}}\right)}{(0.92)\left(1000\ \frac{\text{kg}}{\text{m}^3}\right)\left(9.81\ \frac{\text{m}}{\text{s}^2}\right)}\right)}$$

$$= 6.02 \times 10^{-4}\ \text{m}^3/\text{s} \quad (6 \times 10^{-4}\ \text{m}^3/\text{s})$$

The answer is B.

Problem 27

Water flows through a venturi meter, as shown.

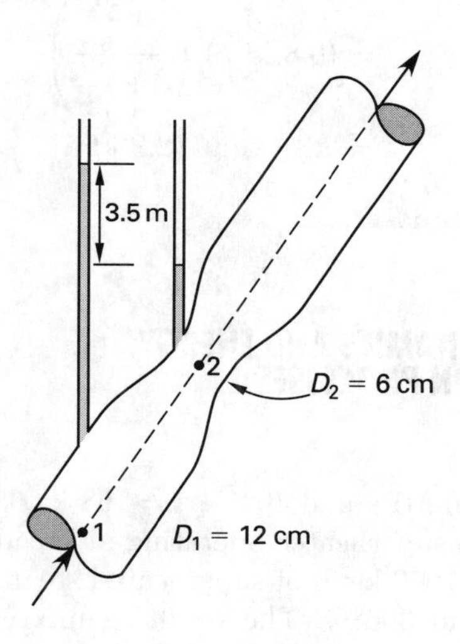

If the static pressure difference between sections 1 and 2 is 3.5 m of water and the coefficient of velocity is 0.98, the volumetric flow rate is most nearly

(A) 0.012 m^3/s
(B) 0.018 m^3/s
(C) 0.024 m^3/s
(D) 0.030 m^3/s

Solution

$$Q = \frac{C_v A_2}{\sqrt{1 - \left(\frac{A_2}{A_1}\right)^2}} \sqrt{2g\left(\frac{\Delta p}{\rho g}\right)}$$

$$= C_v\left(\frac{\pi}{4}\right) D_2^2 \sqrt{\frac{2g\left(\frac{\rho g \Delta h}{\rho g}\right)}{1 - \left(\frac{D_2}{D_1}\right)^4}}$$

$$= (0.98)\left(\frac{\pi}{4}\right)(0.06\ \text{m})^2 \sqrt{\frac{(2)\left(9.81\ \frac{\text{m}}{\text{s}^2}\right)(3.5\ \text{m})}{1 - \left(\frac{0.06\ \text{m}}{0.12\ \text{m}}\right)^4}}$$

$$= 0.0237\ \text{m}^3/\text{s} \quad (0.024\ \text{m}^3/\text{s})$$

The answer is C.

Problem 28

Water flows upward through a venturi meter, as shown. The differential manometer deflection is 50 cm of liquid of specific gravity 1.3.

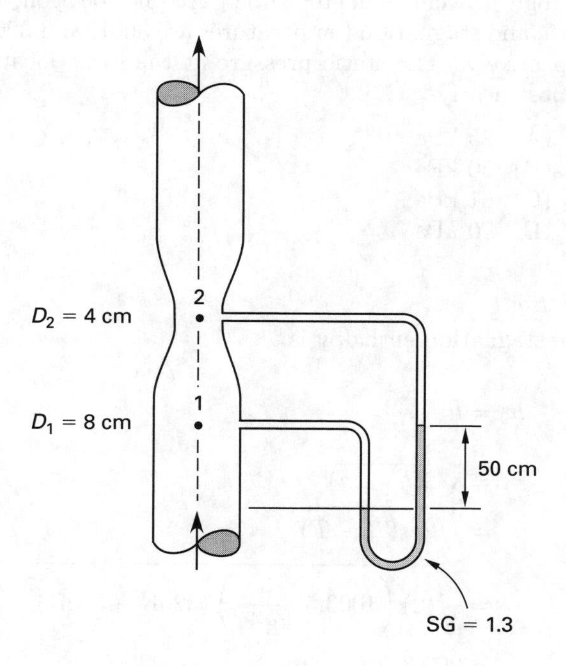

If the coefficient of velocity is 0.95, the flow rate is most nearly

(A) 0.0010 m^3/s
(B) 0.0012 m^3/s
(C) 0.0015 m^3/s
(D) 0.0021 m^3/s

Solution

$$\frac{\Delta p}{\rho_w g} = \frac{(\rho_m - \rho_w)g\Delta h}{\rho_w g}$$
$$= (\text{SG}_m - 1)\Delta h$$
$$= (1.3 - 1)(50\text{ cm})\left(\frac{1\text{ m}}{100\text{ cm}}\right)$$
$$= 0.15\text{ m}$$

$$Q = C_v A_2 \sqrt{\frac{2g\left(\frac{\Delta p}{\rho_w g}\right)}{1 - \left(\frac{A_2}{A_1}\right)^2}}$$
$$= (0.95)\left(\frac{\pi}{4}\right)(0.04\text{ m})^2 \sqrt{\frac{(2)\left(9.81\ \frac{\text{m}}{\text{s}^2}\right)(0.15\text{ m})}{1 - \left(\frac{4\text{ cm}}{8\text{ cm}}\right)^4}}$$
$$= 0.00212\text{ m}^3/\text{s}\quad (0.0021\text{ m}^3/\text{s})$$

The answer is D.

Problem 29

Air flows adiabatically at the rate of 0.1 kg/s through a 3 cm diameter tube. At one location, the static and stagnation temperatures are 300K and 320K, respectively. The static pressure at the same location is most nearly

(A) 25 kPa
(B) 50 kPa
(C) 61 kPa
(D) 70 kPa

Solution

The stagnation enthalpy is

$$h_0 = h + \frac{\text{v}^2}{2}$$
$$\text{v} = \sqrt{2(h_0 - h)}$$
$$= \sqrt{2c_p(T_0 - T)}$$
$$= \sqrt{(2)\left(1003.5\ \frac{\text{J}}{\text{kg}\cdot\text{K}}\right)(320\text{K} - 300\text{K})}$$
$$= 200.3\text{ m/s}$$
$$\rho = \frac{\dot{m}}{A\text{v}}$$
$$= \frac{0.1\ \frac{\text{kg}}{\text{s}}}{\left(\frac{\pi}{4}\right)(0.03\text{ m})^2\left(200.3\ \frac{\text{m}}{\text{s}}\right)}$$
$$= 0.7063\text{ kg/m}^3$$

$$p = \rho RT$$
$$= \left(0.7063\ \frac{\text{kg}}{\text{m}^3}\right)\left(0.287\ \frac{\text{kJ}}{\text{kg}\cdot\text{K}}\right)(300\text{K})$$
$$= 60.8\text{ kPa}\quad (61\text{ kPa})$$

The answer is C.

Problem 30

An error of 0.8 kPa was discovered in reading the velocity head of a liquid flowing at a velocity of 1.6 m/s. What would be the corresponding error in the pressure if the velocity were 3 m/s?

(A) 1.6 kPa
(B) 2.0 kPa
(C) 2.4 kPa
(D) 2.8 kPa

Solution

The velocity head and, therefore, the velocity head error, E, are proportional to the square of the velocity.

$$\frac{E_2}{E_1} = \frac{\text{v}_2^2}{\text{v}_1^2}$$
$$E_2 = E_1\left(\frac{\text{v}_2}{\text{v}_1}\right)^2$$
$$= (0.8\text{ kPa})\left(\frac{3\ \frac{\text{m}}{\text{s}}}{1.6\ \frac{\text{m}}{\text{s}}}\right)^2$$
$$= 2.81\text{ kPa}\quad (2.8\text{ kPa})$$

The answer is D.

THERMODYNAMICS AND ENERGY CONVERSION PROCESSES

Problem 31

Water at 1.0 MPa and 40°C ($h = 168$ kJ/kg) enters an adiabatic desuperheater operating at steady state and mixes with 1000 kg/h of superheated steam entering at 1.0 MPa and 300°C. The resulting mixture leaves as saturated steam at 0.8 MPa.

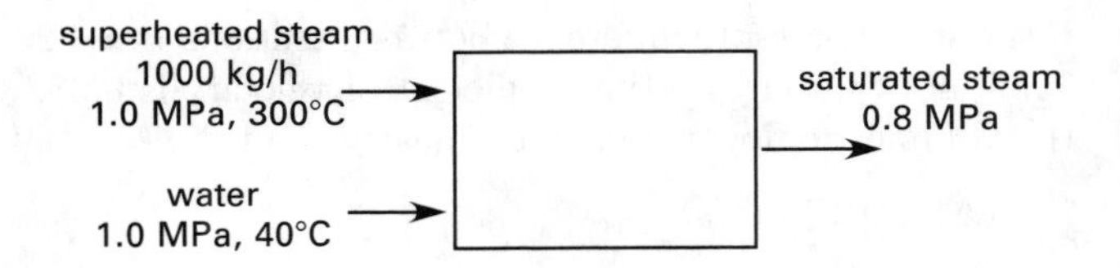

Assuming changes in kinetic and potential energies are negligible, the mass flow rate of the water is most nearly

(A) 50 kg/h
(B) 80 kg/h
(C) 110 kg/h
(D) 140 kg/h

Solution

The continuity equation is

$$\dot{m}_w + \dot{m}_{\text{sh}} = \dot{m}_s$$

(Subscripts w, sh, and s refer to water, superheated steam, and saturated steam, respectively.)

For this problem Q_{in}, W_{in}, KE, and PE are zero, so

$$\sum \dot{m}_i h_i = \sum \dot{m}_e h_e$$

Combining the continuity and energy equations,

$$\dot{m}_w h_w + \dot{m}_{\text{sh}} h_{\text{sh}} = (\dot{m}_{\text{sh}} + \dot{m}_w) h_s$$

Substituting values from steam tables gives

$$\begin{aligned} \dot{m}_w \left(168\ \frac{\text{kJ}}{\text{kg}}\right) + \left(1000\ \frac{\text{kg}}{\text{h}}\right)\left(3051.2\ \frac{\text{kJ}}{\text{kg}}\right) &= \left(\dot{m}_w + 1000\ \frac{\text{kg}}{\text{h}}\right)\left(2769.1\ \frac{\text{kJ}}{\text{kg}}\right) \\ \dot{m}_w &= 108.5\ \text{kg/h} \quad (110\ \text{kg/h}) \end{aligned}$$

The answer is C.

Problem 32

A Carnot engine using air as a working fluid develops 5 kW of power. The engine operates between two thermal reservoirs at 800K and 300K. If the volume doubles during the heat transfer to the engine, the mass flow rate of the air is most nearly

(A) 0.020 kg/s
(B) 0.050 kg/s
(C) 0.10 kg/s
(D) 0.13 kg/s

Solution

$$\begin{aligned} \Delta s &= R \ln \frac{v_2}{v_1} \\ &= \left(0.287\ \frac{\text{kJ}}{\text{kg·K}}\right) \ln 2 \\ &= 0.1989\ \text{kJ/kg·K} \end{aligned}$$

The power output is

$$\begin{aligned} P &= \dot{m}Q \\ &= \dot{m}\Delta T \Delta s \\ \dot{m} &= \frac{P}{\Delta T \Delta s} \\ &= \frac{5\ \text{kW}}{(800\text{K} - 300\text{K})\left(0.1989\ \frac{\text{kJ}}{\text{kg·K}}\right)} \\ &= 0.0503\ \text{kg/s} \quad (0.050\ \text{kg/s}) \end{aligned}$$

The answer is B.

Problem 33

1 kg of water is cooled from 90°C to the surrounding temperature of 20°C. Assuming the specific heat of water is 4.18 kJ/kg·K, the total change in entropy (system and surroundings) is most nearly

(A) 0.05 kJ/K
(B) 0.1 kJ/K
(C) 0.9 kJ/K
(D) 1 kJ/K

Solution

$$\begin{aligned} Q &= mc\Delta T \\ &= (1\ \text{kg})\left(4.18\ \frac{\text{kJ}}{\text{kg·K}}\right)(20^\circ\text{C} - 90^\circ\text{C}) \\ &= -292.6\ \text{kJ} \\ \Delta S_{\text{system}} &= mc \ln \frac{T_2}{T_1} \\ &= (1\ \text{kg})\left(4.18\ \frac{\text{kJ}}{\text{kg·K}}\right) \ln \frac{20^\circ\text{C} + 273^\circ}{90^\circ\text{C} + 273^\circ} \\ &= -0.8955\ \text{kJ/K} \\ \Delta S_\infty &= \frac{Q_\infty}{T_\infty} = \frac{292.6\ \text{kJ}}{20^\circ\text{C} + 273^\circ} \\ &= 0.9986\ \text{kJ/K} \\ \Delta S_{\text{total}} &= -0.8955\ \frac{\text{kJ}}{\text{K}} + 0.9986\ \frac{\text{kJ}}{\text{K}} \\ &= 0.1031\ \text{kJ/K} \quad (0.1\ \text{kJ/K}) \end{aligned}$$

The answer is B.

Problem 34

Air at 1 MPa and 400K expands through an adiabatic turbine operating at steady state to 0.15 MPa and 320K. Changes in kinetic and potential energies are negligible, and air can be assumed to be an ideal gas with constant specific heats. The surrounding temperature is 298.15K.

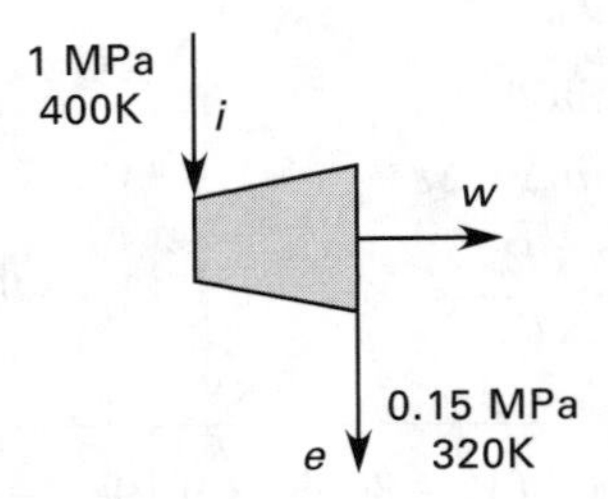

The reversible work that could have been developed if the expansion took place with the same inlet and exit states is most nearly

(A) 120 kJ/kg
(B) 150 kJ/kg
(C) 180 kJ/kg
(D) 210 kJ/kg

Solution

$$\begin{aligned} w_{\text{rev}} &= (h_e - h_i) - T_0(s_e - s_i) \\ &= c_p(T_e - T_i) - T_0\left(c_p \ln \frac{T_e}{T_i} - R \ln \frac{p_e}{p_i}\right) \\ &= \left(1.0035\ \frac{\text{kJ}}{\text{kg·K}}\right)(320\text{K} - 400\text{K}) \\ &\quad - (298.15\text{K})\left(\begin{array}{c} \left(1.0035\ \frac{\text{kJ}}{\text{kg·K}}\right) \ln \frac{320\text{K}}{400\text{K}} \\ - \left(0.287\ \frac{\text{kJ}}{\text{kg·K}}\right) \\ \times \ln \frac{0.15\ \text{MPa}}{1\ \text{MPa}} \end{array}\right) \\ &= -175.85\ \text{kJ/kg} \quad (180\ \text{kJ/kg}) \end{aligned}$$

The answer is C.

Problem 35

Compressed air in a tank is at a pressure of 800 kPa and a temperature of 600K. Assume the air acts as an ideal gas with constant specific heats and the temperature of the environment is 25°C. The closed-system availability exhausting to a standard atmospheric pressure is most nearly

(A) 85 kJ/kg
(B) 100 kJ/kg
(C) 110 kJ/kg
(D) 120 kJ/kg

Solution

$$\begin{aligned} v &= \frac{RT}{p} \\ &= \frac{\left(0.287\ \frac{\text{kJ}}{\text{kg·K}}\right)(600\text{K})}{800\ \text{kPa}} \\ &= 0.21525\ \text{m}^3/\text{kg} \end{aligned}$$

$$\begin{aligned} T_0 &= 25°\text{C} + 273° \\ &= 298\text{K} \end{aligned}$$

$$\begin{aligned} v_0 &= \frac{RT_0}{p_0} \\ &= \frac{\left(0.287\ \frac{\text{kJ}}{\text{kg·K}}\right)(298\text{K})}{101.3\ \text{kPa}} \\ &= 0.8443\ \text{m}^3/\text{kg} \end{aligned}$$

The availability is

$$\begin{aligned} \phi - \phi_0 &= (u - u_0) + p_0(v - v_0) - T_0(s - s_0) \\ &= c_v(T - T_0) + p_0(v - v_0) \\ &\quad - T_0\left(c_p \ln \frac{T}{T_0} - R \ln \frac{p}{p_0}\right) \\ &= \left(0.7165\ \frac{\text{kJ}}{\text{kg·K}}\right)(600\text{K} - 298\text{K}) \\ &\quad + (101.3\ \text{kPa})\left(0.21525\ \frac{\text{m}^3}{\text{kg}} - 0.8443\ \frac{\text{m}^3}{\text{kg}}\right) \\ &\quad - (298\text{K})\left(\begin{array}{c} \left(1.0035\ \frac{\text{kJ}}{\text{kg·K}}\right) \ln \frac{600\text{K}}{298\text{K}} \\ - \left(0.287\ \frac{\text{kJ}}{\text{kg·K}}\right) \\ \times \ln \frac{800\ \text{kPa}}{101.3\ \text{kPa}} \end{array}\right) \\ &= 120.1\ \text{kJ/kg} \quad (120\ \text{kJ/kg}) \end{aligned}$$

The answer is D.

Problem 36

Benzene (C_6H_6) is burned with 20% excess air. The air-fuel ratio by mass is nearest to

(A) 12
(B) 16
(C) 18
(D) 20

Solution

The stoichiometric reaction per volume of benzene is

$$C_6H_6(\ell) + 7.5O_2(g) + (7.5)(3.76)N_2(g) \longrightarrow \\ 6CO_2(g) + 3H_2O(g) + (7.5)(3.76)N_2(g)$$

With 20% excess air,

$$C_6H_6(\ell) + (1.2)(7.5)O_2(g) \\ + (1.2)(7.5)(3.76)N_2(g) \longrightarrow \\ 6CO_2(g) + 3H_2O(g) + (1.2)(7.5)(3.76)N_2(g) \\ + (0.2)(7.5)O_2(g)$$

$$C_6H_6(\ell) + 9O_2(g) + 33.84N_2(g) \longrightarrow \\ 6CO_2(g) + 3H_2O(g) + 33.84N_2(g) + 1.5O_2(g)$$

The molecular weight of benzene is

$$(6)\left(12\ \frac{\text{kg C}}{\text{kmol}}\right) + (6)\left(1\ \frac{\text{kg H}}{\text{mol}}\right) = 78\ \text{kg C}_6\text{H}_6/\text{kmol}$$

The mass air-fuel ratio is

$$\frac{A}{F} = \frac{(9\ \text{kmol})\left(32\ \frac{\text{kg O}_2}{\text{kmol}}\right) + (33.84\ \text{kmol})\left(28\ \frac{\text{kg N}_2}{\text{kmol}}\right)}{78\ \frac{\text{kg C}_6\text{H}_6}{\text{kmol}}} \\ = 15.84\ \text{kg of air/kg of fuel} \quad (16)$$

The answer is B.

Problem 37

The enthalpy values at key points of an ideal reheat steam cycle are shown.

state	enthalpy, h (kJ/kg)
1	3138.3
2	2554.1
3	3271.9
4	2470.9
5	191.83
6	199.9

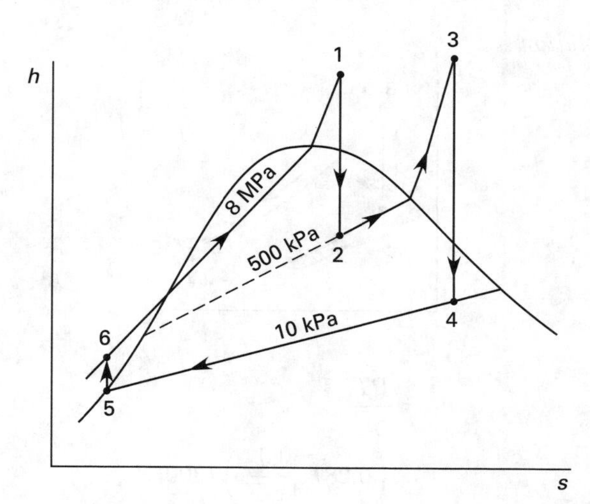

The thermal efficiency of the cycle is most nearly

(A) 33%
(B) 38%
(C) 40%
(D) 42%

Solution

$$\eta_t = \frac{w_{\text{net out}}}{q_{\text{input}}} = \frac{(h_1 - h_2) + (h_3 - h_4) - (h_6 - h_5)}{(h_1 - h_6) + (h_3 - h_2)}$$

$$= \frac{\left(3138.3\ \frac{\text{kJ}}{\text{kg}} - 2554.1\ \frac{\text{kJ}}{\text{kg}}\right) + \left(3271.9\ \frac{\text{kJ}}{\text{kg}} - 2470.9\ \frac{\text{kJ}}{\text{kg}}\right) - \left(199.9\ \frac{\text{kJ}}{\text{kg}} - 191.83\ \frac{\text{kJ}}{\text{kg}}\right)}{\left(3138.3\ \frac{\text{kJ}}{\text{kg}} - 199.9\ \frac{\text{kJ}}{\text{kg}}\right) + \left(3271.9\ \frac{\text{kJ}}{\text{kg}} - 2554.1\ \frac{\text{kJ}}{\text{kg}}\right)} \\ = 0.3767 \quad (38\%)$$

The answer is B.

Problem 38

An air-standard Otto cycle has a compression ratio of 8.0. At the beginning of the compression stroke the temperature is 300K and the pressure is 100 kPa. The maximum temperature in the cycle is 1000K. Assuming constant specific heats, the mean effective pressure of the cycle is most nearly

(A) 140 kPa
(B) 170 kPa
(C) 200 kPa
(D) 220 kPa

Solution

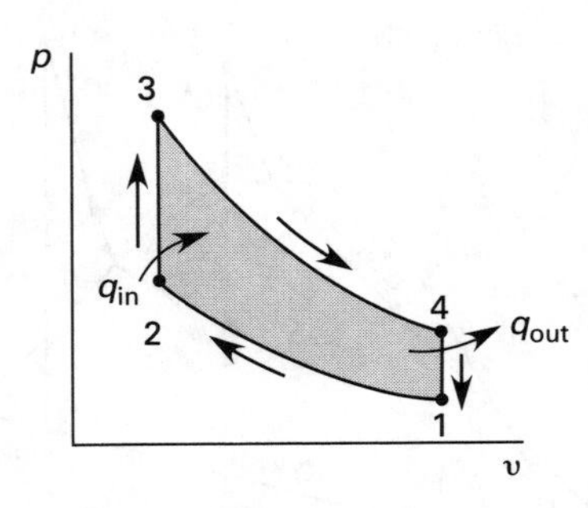

$$v_1 = \frac{RT_1}{p_1} = \frac{\left(0.287\ \frac{\text{kJ}}{\text{kg·K}}\right)(300\text{K})}{100\ \text{kPa}} = 0.861\ \text{m}^3/\text{kg}$$

The compression ratio is a ratio of volumes.

$$r_c = \frac{v_1}{v_2}$$

$$v_2 = \frac{v_1}{r_c} = \frac{0.861\ \frac{\text{m}^3}{\text{kg}}}{8.0} = 0.1076\ \text{m}^3/\text{kg}$$

$$\frac{T_2}{T_1} = \left(\frac{v_1}{v_2}\right)^{k-1} = r_c^{k-1}$$

$$T_2 = (300\text{K})(8.0)^{1.4-1} = 689.2\text{K}$$

$$q_{2-3} + w_{2-3} = u_3 - u_2 \quad [w_{2-3} = 0]$$

$$q_{2-3} = u_3 - u_2 = c_v(T_3 - T_2) = \left(0.7165\ \frac{\text{kJ}}{\text{kg·K}}\right)(1000\text{K} - 689.2\text{K}) = 222.7\ \text{kJ/kg}$$

$$\frac{T_4}{T_3} = \left(\frac{v_3}{v_4}\right)^{k-1} = \left(\frac{1}{r_c}\right)^{k-1}$$

$$T_4 = (1000\text{K})\left(\frac{1}{8.0}\right)^{1.4-1} = 435.3\text{K}$$

$$q_{4-1} = u_1 - u_4 = c_v(T_1 - T_4) = \left(0.7165\ \frac{\text{kJ}}{\text{kg·K}}\right)(300\text{K} - 435.3\text{K}) = -96.94\ \text{kJ/kg}$$

The mean effective pressure is

$$\text{MEP} = \frac{|w_{\text{net out}}|}{v_1 - v_2} = \frac{|q_{in} - q_{out}|}{v_1 - v_2} = \frac{222.7\ \frac{\text{kJ}}{\text{kg}} - 96.94\ \frac{\text{kJ}}{\text{kg}}}{0.861\ \frac{\text{m}^3}{\text{kg}} - 0.1076\ \frac{\text{m}^3}{\text{kg}}} = 166.9\ \text{kPa} \quad (170\ \text{kPa})$$

The answer is B.

Problem 39

The components of a gas turbine operating at steady state are shown. The air enters the compressor at 300K and 100 kPa and is compressed to 800 kPa. The compressor's isentropic efficiency is 80%.

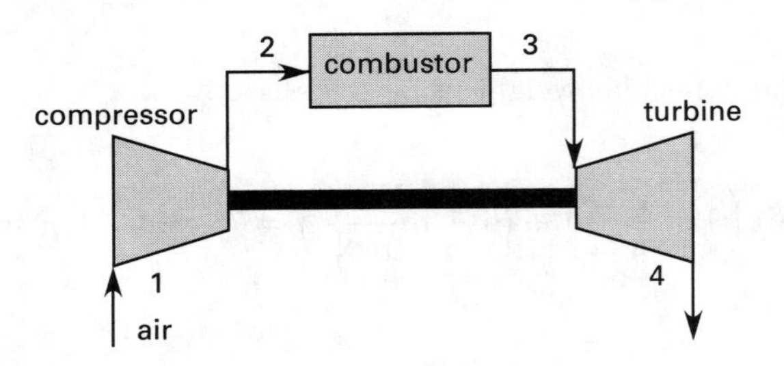

Assuming air to be an ideal gas with constant specific heats, the work required by the compressor is most nearly

(A) 300 kJ/kg
(B) 325 kJ/kg
(C) 330 kJ/kg
(D) 340 kJ/kg

Solution

$$\frac{T_{2s}}{T_1} = \left(\frac{p_2}{p_1}\right)^{\frac{k-1}{k}}$$

$$T_{2s} = (300\text{K})\left(\frac{800\ \text{kPa}}{100\ \text{kPa}}\right)^{\frac{1.4-1}{1.4}} = 543.4\text{K}$$

$$\eta = \frac{W_{\text{isentropic}}}{W_{\text{actual}}} = \frac{h_{2s} - h_1}{h_2 - h_1}$$

$$W_{\text{actual}} = h_2 - h_1 = \frac{h_{2s} - h_1}{\eta} = \frac{c_p(T_{2s} - T_1)}{\eta} = \frac{\left(1.0035\ \frac{\text{kJ}}{\text{kg·K}}\right)(543.4\text{K} - 300\text{K})}{0.8} = 305.3\ \text{kJ/kg} \quad (300\ \text{kJ/kg})$$

The answer is A.

FLUID MECHANICS AND FLUID MACHINERY

Problem 40

A sphere, 10 cm in diameter, floats in 20°C water with half of its volume submerged. The density of water at 20°C is 998 kg/m^3. The mass of the sphere is most nearly

(A) 0.20 kg
(B) 0.26 kg
(C) 0.30 kg
(D) 2.6 kg

Solution

The buoyant force is equal to the weight of the sphere.

$$\begin{aligned} W &= F_b = mg \\ &= \rho_{\text{water}} V g \end{aligned}$$

The submerged volume is

$$\begin{aligned} V &= \tfrac{1}{2} V_{\text{sphere}} \\ &= \left(\frac{1}{2}\right)\left(\frac{\pi}{6}\right) D^3 \\ &= \left(\frac{1}{2}\right)\left(\frac{\pi}{6}\right) (0.1\ \text{m})^3 \\ &= 0.2618 \times 10^{-3}\ \text{m}^3 \\ m &= \rho_{\text{water}} V \\ &= \left(998\ \frac{\text{kg}}{\text{m}^3}\right)(0.2618 \times 10^{-3}\ \text{m}^3) \\ &= 0.261\ \text{kg} \quad (0.26\ \text{kg}) \end{aligned}$$

The answer is B.

Problem 41

A rectangular gate 0.5 m wide is located in a fresh water tank at a slope of 45°C, as shown. The gate is hinged along the top edge and is held in place by a force F at the bottom edge.

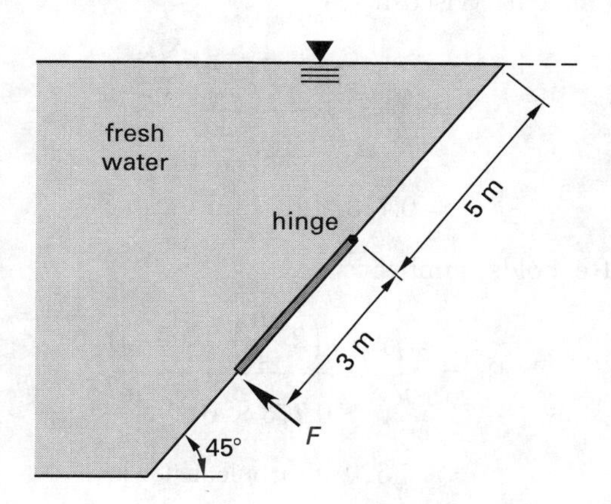

Neglecting the weight of the gate and any friction at the hinge, the force F is most nearly

(A) 21 kN
(B) 32 kN
(C) 36 kN
(D) 43 kN

Solution

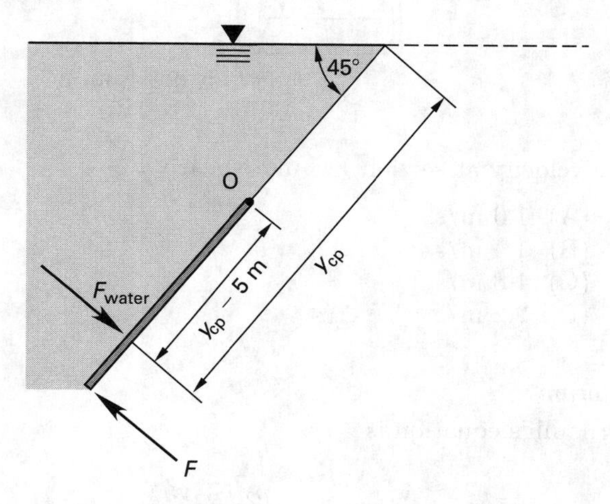

$$\begin{aligned} F_{\text{water}} &= \rho g h_{\text{cg}} A \\ &= \left(1000\ \frac{\text{kg}}{\text{m}^3}\right)\left(9.81\ \frac{\text{m}}{\text{s}^2}\right)(6.5\ \text{m}) \sin 45^\circ \\ &\quad \times (0.5\ \text{m})(3\ \text{m}) \\ &= 67\,633\ \text{N} \\ y_{\text{cg}} &= 5\ \text{m} + \frac{3\ \text{m}}{2} \\ &= 6.5\ \text{m} \end{aligned}$$

The location of this force is

$$\begin{aligned} y_{\text{cp}} &= \frac{I_{\text{cg}}}{y_{\text{cg}} A} + y_{\text{cg}} \\ &= \frac{\frac{1}{12} b h^3}{y_{\text{cg}} b h} + y_{\text{cg}} \\ &= \frac{\left(\frac{1}{12}\right)(0.5\ \text{m})(3\ \text{m})^3}{(6.5\ \text{m})(0.5\ \text{m})(3\ \text{m})} + 6.5\ \text{m} \\ &= 6.6154\ \text{m} \end{aligned}$$

$$\begin{aligned} \sum M_{\text{hinge}} &= 0 \\ F_{\text{water}}(y_{\text{cp}} - 5\ \text{m}) &= F(3\ \text{m}) \\ F &= \frac{(67\,633\ \text{N})(6.6154\ \text{m} - 5\ \text{m})}{3\ \text{m}} \\ &= 36\,418\ \text{N} \quad (36\ \text{kN}) \end{aligned}$$

The answer is C.

Problem 42

Water flows steadily through the contraction shown.

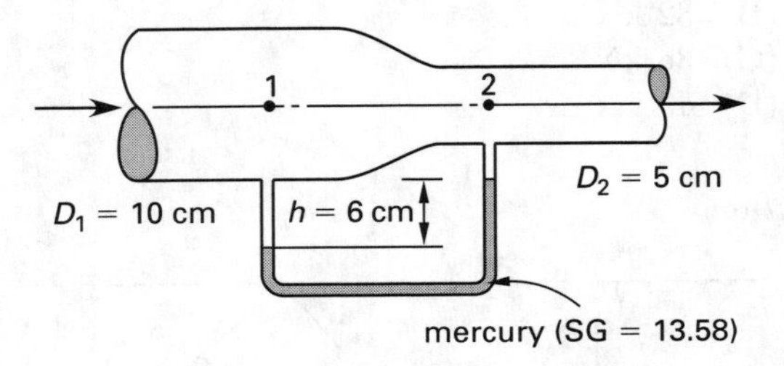

The velocity at section 1 is most nearly

(A) 1.0 m/s
(B) 1.4 m/s
(C) 1.8 m/s
(D) 2.2 m/s

Solution

Bernoulli's equation is

$$\frac{p_1}{\rho_w g} + \frac{v_1^2}{2g} + z_1 = \frac{p_2}{\rho_w g} + \frac{v_2^2}{2g} + z_2$$

$$z_1 = z_2$$

$$v_2 = \left(\frac{D_1}{D_2}\right)^2 v_1 = \left(\frac{10 \text{ cm}}{5 \text{ cm}}\right)^2 v_1$$
$$= 4v_1$$

Substituting into Bernoulli's equation,

$$\frac{p_1 - p_2}{\rho_w g} = \frac{16v_1^2 - v_1^2}{2g} = \frac{15v_1^2}{2g}$$

$$p_1 - p_2 = hg(\rho_{\text{Hg}} - \rho_w)$$

$$\frac{p_1 - p_2}{\rho_w g} = \left(\frac{\rho_{\text{Hg}}}{\rho_w} - 1\right) h$$

$$\frac{15v_1^2}{2g} = \left(\frac{\rho_{\text{Hg}}}{\rho_w} - 1\right) h$$

$$v_1 = \sqrt{\left(\frac{2g}{15}\right)\left(\frac{\rho_{\text{Hg}}}{\rho_w} - 1\right) h}$$

$$= \sqrt{\left(\frac{(2)\left(9.81 \ \frac{\text{m}}{\text{s}^2}\right)}{15}\right)(13.58 - 1)(0.06 \text{ m})}$$

$$= 0.9936 \text{ m/s} \quad (1.0 \text{ m/s})$$

The answer is A.

Problem 43

Water flows in an inclined constant-diameter pipe. At point 1, $p_1 = 235$ kPa, and the elevation is $z_1 = 20$ m. At point 2, $p_2 = 200$ kPa, and $z_2 = 22$ m. The friction head loss between the two sections is most nearly

(A) 0.80 m
(B) 1.2 m
(C) 1.6 m
(D) 1.9 m

Solution

$$\frac{p_1}{\rho g} + \frac{v_1^2}{2g} + z_1 = \frac{p_2}{\rho g} + \frac{v_2^2}{2g} + z_2 + h_{f_{1-2}}$$

$$v_1 = v_2$$

$$h_{f_{1-2}} = \frac{p_1 - p_2}{\rho g} + z_1 - z_2$$

$$= \frac{(235 \text{ kPa} - 200 \text{ kPa})\left(1000 \ \frac{\text{Pa}}{\text{kPa}}\right)}{\left(1000 \ \frac{\text{kg}}{\text{m}^3}\right)\left(9.81 \ \frac{\text{m}}{\text{s}^2}\right)}$$
$$+ 20 \text{ m} - 22 \text{ m}$$
$$= 1.568 \text{ m} \quad (1.6 \text{ m})$$

The answer is C.

Problem 44

Water at 32°C flows at 2 m/s in a pipe having an inside diameter of 3 cm. The viscosity of the water is 769×10^{-6} N·s/m^2, and the density is 995 kg/m^3. If the relative roughness of the pipe is 0.002, the friction factor is most nearly

(A) 0.025
(B) 0.030
(C) 0.035
(D) 0.040

Solution

The kinematic viscosity is

$$\nu = \frac{\mu}{\rho} = \frac{769 \times 10^{-6} \ \frac{\text{N·s}}{\text{m}^2}}{995 \ \frac{\text{kg}}{\text{m}^3}}$$
$$= 0.773 \times 10^{-6} \text{ m}^2/\text{s}$$

The Reynolds number is

$$\text{Re} = \frac{vD}{\nu} = \frac{\left(2 \ \frac{\text{m}}{\text{s}}\right)(0.03 \text{ m})}{0.773 \times 10^{-6} \ \frac{\text{m}^2}{\text{s}}}$$
$$= 77\,620 \quad \text{[turbulent flow]}$$

From the Moody diagram at Re = 77620 and ε/D = 0.002, the friction factor is

$$f = 0.0254 \quad (0.025)$$

The answer is A.

Problem 45

An airplane is traveling at 1900 km/h at an altitude where the temperature is −60°C. The Mach number at which the airplane is flying is most nearly

(A) 0.80
(B) 1.3
(C) 1.6
(D) 1.8

Solution

$$\begin{aligned} a = \sqrt{kRT} &= \sqrt{k\left(\frac{\overline{R}}{\text{MW}}\right)T} \\ &= \sqrt{(1.4)\left(\frac{8314\ \frac{\text{J}}{\text{kmol·K}}}{29\ \frac{\text{kg}}{\text{kmol}}}\right)(-60°\text{C} + 273)} \\ &= 292.4\ \text{m/s} \end{aligned}$$

$$\begin{aligned} \text{M} &= \frac{\text{v}}{a} \\ &= \frac{\left(1900\ \frac{\text{km}}{\text{h}}\right)\left(1000\ \frac{\text{m}}{\text{km}}\right)\left(\frac{1\ \text{h}}{3600\ \text{s}}\right)}{292.4\ \frac{\text{m}}{\text{s}}} \\ &= 1.8 \end{aligned}$$

The answer is D.

Problem 46

A pump is used to draw water from a lake, as shown. The friction head loss in the intake line is 1.5 m of water. The pump is located at an elevation of 6 m from the surface, as indicated. The water temperature is 20°C. The vapor pressure of 20°C water is 2.339 kPa.

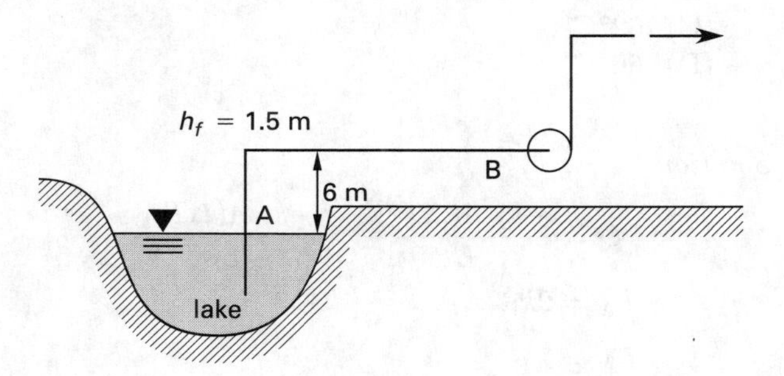

The net positive suction head available is most nearly

(A) 2.6 m of water
(B) 4.5 m of water
(C) 7.5 m of water
(D) 11 m of water

Solution

The atmospheric head is

$$\begin{aligned} \frac{p_{\text{A}}}{\rho g} &= \frac{(101.3\ \text{kPa})\left(1000\ \frac{\text{Pa}}{\text{kPa}}\right)}{\left(1000\ \frac{\text{kg}}{\text{m}^3}\right)\left(9.81\ \frac{\text{m}}{\text{s}^2}\right)} \\ &= 10.3\ \text{m}\ \text{H}_2\text{O} \end{aligned}$$

At 20°C, the vapor pressure head is

$$\begin{aligned} \frac{p_{\text{v}}}{\rho g} &= \frac{(2.339\ \text{kPa})\left(1000\ \frac{\text{Pa}}{\text{kPa}}\right)}{\left(1000\ \frac{\text{kg}}{\text{m}^3}\right)\left(9.81\ \frac{\text{m}}{\text{s}^2}\right)} \\ &= 0.24\ \text{m}\ \text{H}_2\text{O} \end{aligned}$$

The net positive suction head available is

$$\begin{aligned} \text{NPSHA} &= \frac{p_{\text{A}}}{\rho g} - z_{\text{B}} - h_f - \frac{p_{\text{v}}}{\rho g} \\ &= 10.3\ \text{m} - 6\ \text{m} - 1.5\ \text{m} - 0.24\ \text{m} \\ &= 2.56\ \text{m} \quad (2.6\ \text{m}) \end{aligned}$$

The answer is A.

Problem 47

The system characteristic curve for a centrifugal pump is shown. 20°C water flows through 100 m of 20 cm diameter pipe. The pipe is made of galvanized iron with a friction factor of 0.02.

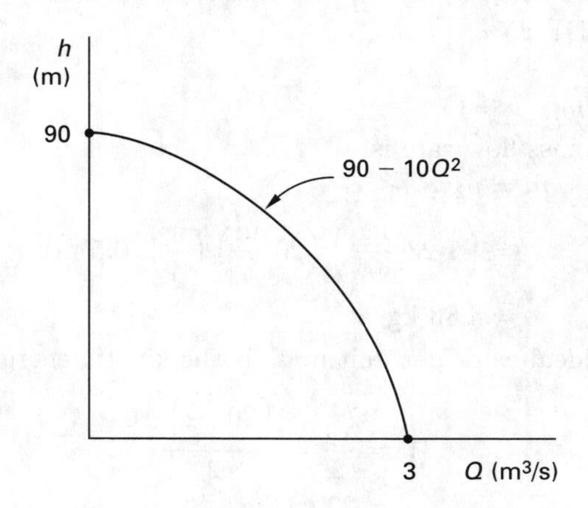

The flow rate is most nearly

(A) $0.4\ \text{m}^3/\text{s}$
(B) $0.7\ \text{m}^3/\text{s}$
(C) $1\ \text{m}^3/\text{s}$
(D) $2\ \text{m}^3/\text{s}$

Solution

$$h_p = h_f$$

$$90 - 10Q^2 = \frac{f L \text{v}^2}{2Dg} = \frac{8fLQ^2}{\pi^2 g D^5}$$

$$= \frac{(8)(0.02)(100\ \text{m})Q^2}{\pi^2 \left(9.81\ \frac{\text{m}}{\text{s}^2}\right)(0.2\ \text{m})^5}$$

$$= 516.4Q^2$$

$$Q = \sqrt{\frac{90\ \text{m}}{10\ \frac{\text{s}^2}{\text{m}^5} + 516.4\ \frac{\text{s}^2}{\text{m}^5}}}$$

$$= 0.413\ \text{m}^3/\text{s} \quad (0.4\ \text{m}^2/\text{s})$$

The answer is A.

Problem 48

A ventilating fan accelerates air with a density of $1.23\ \text{kg/m}^3$ to 20 m/s in a 0.5 m diameter duct. The power input to the fan blades is 1.5 kW.

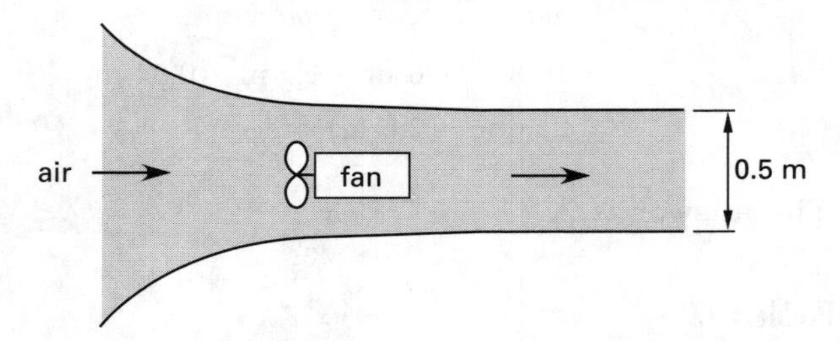

The efficiency of the fan is most nearly

(A) 40%
(B) 65%
(C) 75%
(D) 85%

Solution

The mass flow rate is

$$\dot{m} = \rho_2 \text{v}_2 A_2$$

$$= \left(1.23\ \frac{\text{kg}}{\text{m}^3}\right)\left(20\ \frac{\text{m}}{\text{s}}\right)\left(\frac{\pi}{4}\right)(0.5\ \text{m})^2$$

$$= 4.83\ \text{kg/s}$$

The ideal work per unit mass is the kinetic energy.

$$W = \frac{\text{v}_2^2}{2} = \frac{\left(20\ \frac{\text{m}}{\text{s}}\right)^2}{2}$$

$$= 200\ \text{J/kg}$$

The work performed per unit mass is

$$W_{\text{actual}} = \frac{P}{\dot{m}}$$

$$= \frac{(1.5\ \text{kW})\left(1000\ \frac{\text{W}}{\text{kW}}\right)}{4.83\ \frac{\text{kg}}{\text{s}}}$$

$$= 310.6\ \text{J/kg}$$

$$\eta = \frac{W_{\text{ideal}}}{W_{\text{actual}}} = \frac{200\ \frac{\text{J}}{\text{kg}}}{310.6\ \frac{\text{J}}{\text{kg}}}$$

$$= 0.644 \quad (65\%)$$

The answer is B.

HEAT TRANSFER

Problem 49

Heat flows steadily through a composite wall made up of two materials, A and B, of equal thickness. The thermal conductivity of material A is double that of material B.

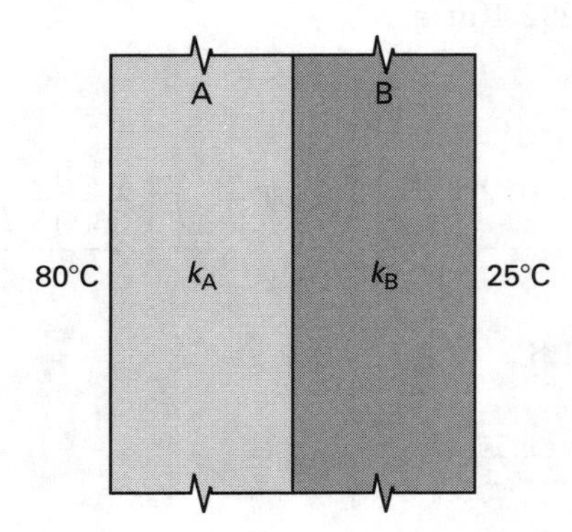

If the temperature of the outside surfaces of A and B are 80°C and 25°C, respectively, the temperature of the contact surface is most nearly

(A) 54°C
(B) 58°C
(C) 62°C
(D) 66°C

Solution

$$Q = \frac{-k_\text{A} A (\Delta T)_\text{A}}{L_\text{A}} = \frac{-k_\text{B} A (\Delta T)_\text{B}}{L_\text{B}}$$

$$k_\text{A} = 2k_\text{B}$$

$$L_\text{A} = L_\text{B}$$

Therefore,

$$(\Delta T)_A = \frac{(\Delta T)_B}{2}$$

$$(2)(80°C - T) = T - 25°C$$

$$T = 61.7°C \quad (62°C)$$

The answer is C.

Problem 50

An 8 m long pipe of 15 cm outside diameter is covered with 2 cm of insulation with thermal conductivity of 0.09 W/m·K.

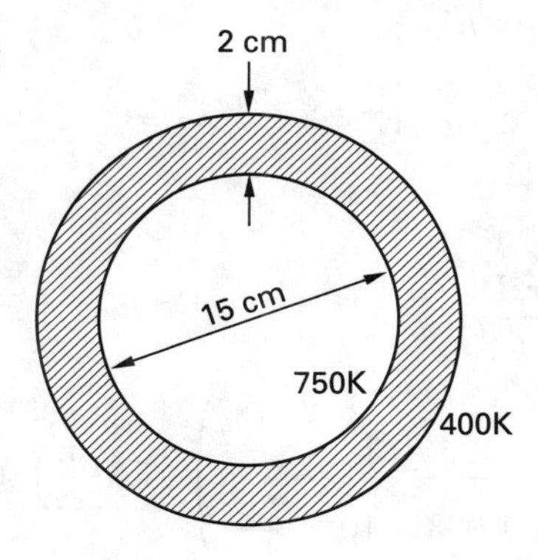

If the inner and outer temperatures of the insulation are 750K and 400K, respectively, what is the heat loss from the pipe?

(A) 4.5 kW
(B) 6.7 kW
(C) 8.5 kW
(D) 10 kW

Solution

$$Q = \frac{2\pi L k (T_1 - T_2)}{\ln \frac{r_2}{r_1}}$$

$$= \frac{2\pi(8 \text{ m})\left(0.09 \frac{\text{W}}{\text{m·K}}\right)(750\text{K} - 400\text{K})}{\ln \frac{9.5 \text{ cm}}{7.5 \text{ cm}}}$$

$$= 6698 \text{ W} \quad (6.7 \text{ kW})$$

The answer is B.

Problem 51

Heat is generated internally at the rate of 1 MW/m^3 in a plane wall of thickness 0.07 m with a thermal conductivity of 18 W/m·K. One side of the wall is insulated, and the other side is exposed to surroundings at 300K.

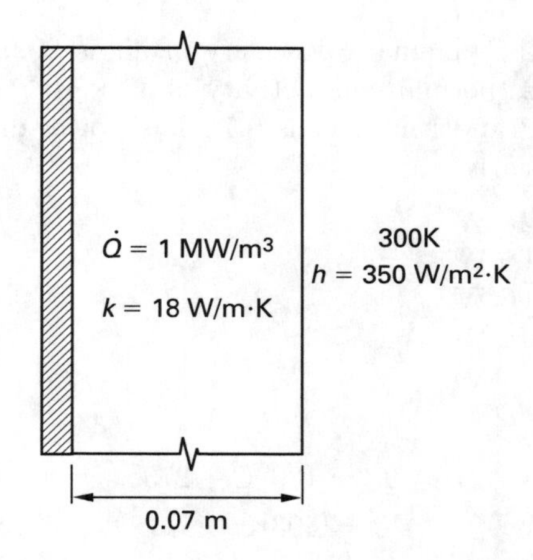

If the convective heat transfer coefficient between the wall and the surroundings is 350 W/m^2·K, the maximum temperature in the wall is likely to be

(A) 500K
(B) 550K
(C) 640K
(D) 700K

Solution

The energy generated is equal to the convective heat loss.

$$\dot{Q}AL = hA(T_s - T_\infty)$$

$$T_s = \frac{\dot{Q}L}{h} + T_\infty$$

$$= \frac{\left(1 \frac{\text{MW}}{\text{m}^3}\right)\left(10^6 \frac{\text{W}}{\text{MW}}\right)(0.07 \text{ m})}{350 \frac{\text{W}}{\text{m}^2\text{·K}}} + 300\text{K}$$

$$= 500\text{K}$$

The maximum temperature occurs at the insulated surface and is given by

$$T_{\text{max}} = T_s + \left(\frac{\dot{Q}}{2k}\right) L^2$$

$$= 500\text{K} + \left(\frac{\left(1 \frac{\text{MW}}{\text{m}^3}\right)\left(10^6 \frac{\text{W}}{1 \text{ MW}}\right)}{(2)\left(18 \frac{\text{W}}{\text{m·K}}\right)}\right)(0.07 \text{ m})^2$$

$$= 636\text{K} \quad (640\text{K})$$

The answer is C.

Problem 52

An electrically heated plate is mounted vertically in 25°C air. The plate has a surface area of 0.1 m^2, has a height of 0.3 m, and is maintained at a uniform temperature of 130°C.

Assume the kinematic viscosity of air is 20.92×10^{-6} m^2/s, the thermal conductivity is 30×10^{-3} W/m·K, and the Prandtl number is 0.7. The power dissipation is most nearly

(A) 45 W
(B) 60 W
(C) 66 W
(D) 74 W

Solution

$$\begin{aligned} T_s &= 130°\text{C} + 273° \\ &= 403\text{K} \\ T_\infty &= 25°\text{C} + 273° \\ &= 298\text{K} \end{aligned}$$

The coefficient of volumetric expansion is the reciprocal of the absolute temperature of the film, which is the average of the surface and local temperatures.

$$\begin{aligned} \beta &= \frac{2}{T_s + T_\infty} \\ &= \frac{2}{403\text{K} + 298\text{K}} \\ &= 0.00285\text{K}^{-1} \end{aligned}$$

The Rayleigh number is

$$\begin{aligned} \text{Ra} &= \frac{g\beta(T_s - T_\infty)L^3\text{Pr}}{\nu^2} \\ &= \frac{\left(9.81\ \frac{\text{m}}{\text{s}^2}\right)(0.00285\text{K}^{-1}) \times (403\text{K} - 298\text{K})(0.3\ \text{m})^3(0.7)}{\left(20.92 \times 10^{-6}\ \frac{\text{m}}{\text{s}^2}\right)^2} \\ &= 1.27 \times 10^8 \end{aligned}$$

At this value of Ra, from tabulated values, $C = 0.59$ and $n = 1/4$. The heat transfer coefficient is

$$\begin{aligned} h &= C\left(\frac{k}{L}\right)\text{Ra}^n \\ &= (0.59)\left(\frac{30 \times 10^{-3}\ \frac{\text{W}}{\text{m·K}}}{0.3\ \text{m}}\right)(1.27 \times 10^8)^{\frac{1}{4}} \\ &= 6.26\ \text{W/m}^2\text{·K} \end{aligned}$$

$$\begin{aligned} Q &= hA\Delta T \\ &= \left(6.26\ \frac{\text{W}}{\text{m}^2\text{·K}}\right)(0.1\ \text{m}^2)(403\text{K} - 298\text{K}) \\ &= 65.7\ \text{W} \quad (66\ \text{W}) \end{aligned}$$

The answer is C.

Problem 53

Water at a bulk temperature of 300K flows inside a long hot pipe of 3 cm inside diameter and at a velocity of 1.3 m/s. At 300K, water properties are

$$\begin{aligned} v &= 1.003 \times 10^{-3}\ \text{m}^3/\text{kg} \\ c_p &= 4.178\ \text{kJ/kg·K} \\ \mu &= 855 \times 10^{-6}\ \text{N·s/m}^2 \\ k &= 613 \times 10^{-3}\ \text{W/m·K} \\ \text{Pr} &= 5.83 \end{aligned}$$

The heat transfer coefficient is most nearly

(A) 2 kW/m^2·K
(B) 3 kW/m^2·K
(C) 4 kW/m^2·K
(D) 5 kW/m^2·K

Solution

$$\begin{aligned} \text{Re} &= \frac{\rho \text{v} D}{\mu} \\ &= \frac{\left(\frac{1}{1.003 \times 10^{-3}\ \frac{\text{m}^3}{\text{kg}}}\right)\left(1.3\ \frac{\text{m}}{\text{s}}\right)(0.03\ \text{m})}{855 \times 10^{-6}\ \frac{\text{N·s}}{\text{m}^2}} \\ &= 45\,478 \quad [\text{turbulent flow}] \\ \text{Nu} &= 0.023\ \text{Re}^{0.8}\text{Pr}^{0.4} \\ &= (0.023)(45\,478)^{0.8}(5.83)^{0.4} \\ &= 247.9 \\ h &= \text{Nu}\left(\frac{k}{D}\right) \\ &= (247.9)\left(\frac{613 \times 10^{-3}\ \frac{\text{W}}{\text{m·K}}}{0.03\ \text{m}}\right) \\ &= 5065\ \text{W/m}^2\text{·K} \quad (5\ \text{kW/m}^2\text{·K}) \end{aligned}$$

The answer is D.

Problem 54

A hot 2 cm diameter metal sphere radiates to a low-temperature enclosure. If 12 W of power is needed to maintain the sphere at 1000K, the emissivity of the sphere is most nearly

(A) 0.17
(B) 0.20
(C) 0.25
(D) 0.27

Solution

$$E = \frac{Q}{A} = \frac{12\ \text{W}}{\pi D^2}$$

$$E_b = \sigma T^4$$

$$\varepsilon = \frac{E}{E_b} = \frac{\frac{12\ \text{W}}{\pi(0.02\ \text{m})^2}}{\left(5.67\times 10^{-8}\ \frac{\text{W}}{\text{m}^2\cdot\text{K}^4}\right)(1000\text{K})^4}$$

$$= 0.168\quad(0.17)$$

The answer is A.

REFRIGERATION AND HVAC

Problem 55

An ideal vapor-compression refrigeration cycle uses R-134a as a refrigerant. The cycle operates between 0.1 MPa and 0.7 MPa. If the flow rate of the refrigerant is 0.15 kg/s, the rate of heat transfer in the evaporator is most nearly

(A) 14 kW
(B) 18 kW
(C) 22 kW
(D) 26 kW

Solution

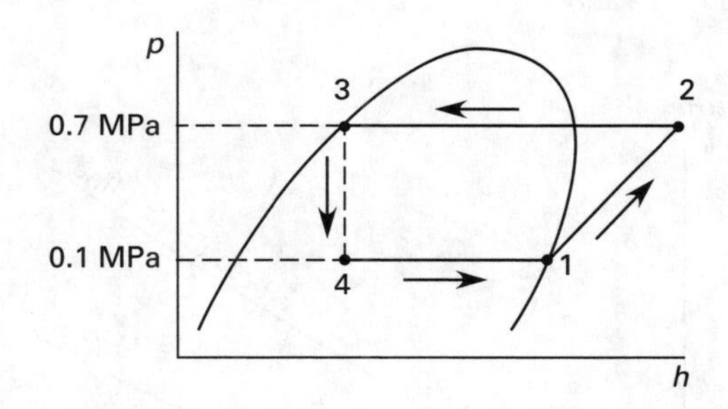

Find the enthalpies at each state using a pressure-enthalpy (*p-h*) diagram or property table for refrigerant HFC-134a.

At $p_1 = 0.1$ MPa, $h_1 = h_{g_1} = 382.8$ kJ/kg. At $p_3 = 0.7$ MPa, $h_3 = h_4 = h_{f_3} = 237.0$ kJ/kg.

$$Q = \dot{m}(h_1 - h_4)$$

$$= \left(0.15\ \frac{\text{kg}}{\text{s}}\right)\left(382.8\ \frac{\text{kJ}}{\text{kg}} - 237.0\ \frac{\text{kJ}}{\text{kg}}\right)$$

$$= 21.87\ \text{kW}\quad(22\ \text{kW})$$

The answer is C.

Problem 56

Air enters the compressor of an ideal-gas refrigeration open cycle at 25°C and 1 atm and is compressed isentropically to 3 atm. The air is then cooled to 75°C before expanding isentropically to 1 atm in a turbine. If the air flow rate is 0.1 kg/s, what is the net power input to the cycle?

(A) 1.2 kW
(B) 1.3 kW
(C) 1.5 kW
(D) 1.6 kW

Solution

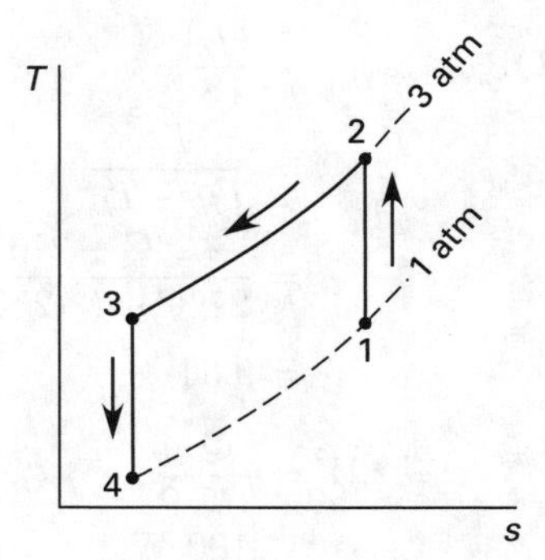

With isentropic compression and expansion in the compressor and turbine,

$$T_2 = T_1\left(\frac{p_2}{p_1}\right)^{\frac{k-1}{k}}$$

$$= (25^\circ\text{C} + 273^\circ)\left(\frac{3\ \text{atm}}{1\ \text{atm}}\right)^{\frac{1.4-1}{1.4}}$$

$$= 407.9\text{K}$$

$$T_4 = T_3\left(\frac{p_4}{p_3}\right)^{\frac{k-1}{k}}$$

$$= (75^\circ\text{C} + 273^\circ)\left(\frac{1\ \text{atm}}{3\ \text{atm}}\right)^{\frac{1.4-1}{1.4}}$$

$$= 254.2\text{K}$$

The net power input is

$$p = \dot{m}\big((h_2 - h_1) - (h_3 - h_4)\big)$$

$$= \dot{m}c_p\big((T_2 - T_1) - (T_3 - T_4)\big)$$

$$= \left(0.1\ \frac{\text{kg}}{\text{s}}\right)\left(1.0035\ \frac{\text{kJ}}{\text{kg}\cdot\text{K}}\right)$$

$$\times (407.9\text{K} - 298\text{K} - 348\text{K} + 254.2\text{K})$$

$$= 1.62\ \text{kW}\quad(1.6\ \text{kW})$$

The answer is D.

Problem 57

A thermoelectric refrigerator removes 100 W of energy from a refrigerated space maintained at −2°C. It rejects energy to an environment at 27°C. The minimum input power required is

(A) 5.0 W
(B) 8.0 W
(C) 11 W
(D) 13 W

Solution

The maximum (ideal) coefficient of performance is

$$\begin{aligned} \text{COP}_{\text{refrigerator}} &= \frac{Q_L}{W} \\ &= \frac{T_L}{T_H - T_L} \\ &= \frac{-2^\circ\text{C} + 273^\circ}{27^\circ\text{C} - (-2^\circ\text{C})} \\ &= 9.34 \end{aligned}$$

$$\begin{aligned} W &= \frac{Q_L}{\text{COP}} \\ &= \frac{100\ \text{W}}{9.34} \\ &= 10.7\ \text{W} \quad (11\ \text{W}) \end{aligned}$$

The answer is C.

Problem 58

The temperature of a glass window in a room is 15°C. If the air temperature in the room is 25°C, the maximum relative humidity before condensation occurs on the glass is most nearly

(A) 35%
(B) 45%
(C) 55%
(D) 65%

Solution

During the sensible cooling process from state 1 to state 2, ω and p_v remain constant.

At state 2, $\phi_2 = 100\%$ so that

$$\begin{aligned} p_{v,1} &= (p_{v,\text{sat}})_{15^\circ\text{C}} \\ &= 1.7051\ \text{kPa} \end{aligned}$$

$$\begin{aligned} \phi_1 &= \frac{p_{v,2}}{(p_{v,\text{sat}})_{25^\circ\text{C}}} \\ &= \frac{1.7051\ \text{kPa}}{3.169\ \text{kPa}} \\ &= 0.538 \quad (55\%) \end{aligned}$$

This can also be found graphically on the psychrometric chart.

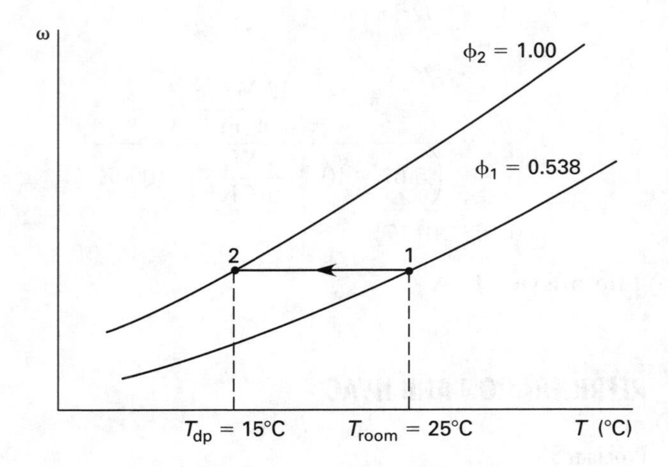

The answer is C.

Problem 59

Outside air at a pressure of 101 kPa, a temperature of 10°C, and a relative humidity of 70% is heated to a temperature of 25°C. If the incoming volumetric flow rate is 1 m^3/s, the rate of heat transfer is most nearly

(A) 8.0 kW
(B) 11 kW
(C) 15 kW
(D) 19 kW

Solution

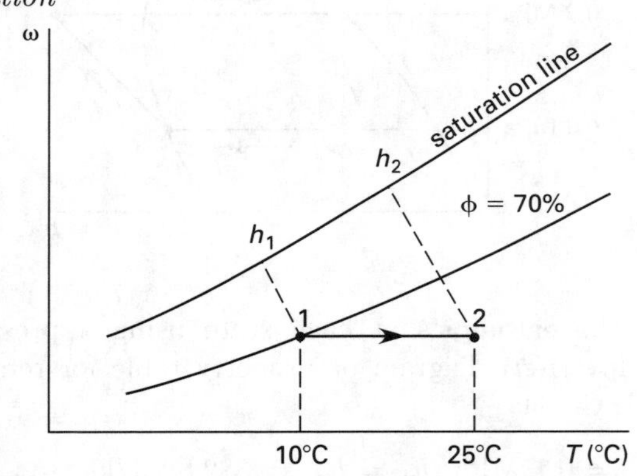

From the psychrometric chart at $T_1 = 10^\circ\text{C}$ and $\phi_1 = 70\%$,

$$h_1 = 23.6\ \text{kJ/kg of dry air}$$

During the heating process the specific humidity, ω, remains constant.

Following a horizontal line to 25°C, determine state 2.

From the chart, $h_2 = 39$ kJ/kg of dry air.

At 10°C,

$$\begin{aligned} p_{v,\text{sat}} &= 1.2276 \text{ kPa} \\ p_v &= \phi p_{v,\text{sat}} = (0.7)(1.2276 \text{ kPa}) \\ &= 0.859 \text{ kPa} \end{aligned}$$

The partial pressure of the dry air is

$$\begin{aligned} p_a &= p_{\text{air}} - p_v = 101 \text{ kPa} - 0.859 \text{ kPa} \\ &= 100.1 \text{ kPa} \end{aligned}$$

The mass flow rate of the dry air is

$$\begin{aligned} \dot{m}_a &= \frac{p_a \dot{V}_a}{R_a T_a} \\ &= \frac{(100.1 \text{ kPa})\left(1.0 \frac{\text{m}^3}{\text{s}}\right)}{\left(0.287 \frac{\text{kJ}}{\text{kg}\cdot\text{K}}\right)(10°\text{C} + 273°)} \\ &= 1.23 \text{ kg/s} \end{aligned}$$

$$\begin{aligned} Q &= \dot{m}_a(h_2 - h_1) \\ &= \left(1.23 \frac{\text{kg}}{\text{s}}\right)\left(39 \frac{\text{kJ}}{\text{kg}} - 23.6 \frac{\text{kJ}}{\text{kg}}\right) \\ &= 18.9 \text{ kW} \quad (19 \text{ kW}) \end{aligned}$$

The answer is D.

Problem 60

Estimate the amount of natural gas required to heat a residence in Billings, Montana using the modified degree day method. The heating value of the fuel is 10 350 W·h/m^3. The calculated heat loss from the house is 23 500 W with indoor and outdoor design temperatures of 21°C and −23°C, respectively. The furnace efficiency factor is approximately 0.55. For Billings, Montana, the degree day, DD, equals 7049°F·d. The correction factor, C, equals 0.64.

The amount of natural gas is most nearly

(A) 5600 m^3
(B) 7500 m^3
(C) 8800 m^3
(D) 10 000 m^3

Solution

Fuel consumption, F, is

$$F = \left(\frac{24(\text{DD})Q}{\eta\,(T_i - T_o)\,H}\right) C$$

The following factors are given in the problem.

$$\begin{aligned} Q &= 23\ 500 \text{ W} \\ \eta &= 0.55 \\ H &= 10\,350 \ \frac{\text{W}\cdot\text{h}}{\text{m}^3} \\ T_i &= 21°\text{C} \\ T_o &= -23°\text{C} \end{aligned}$$

DD is in °F-days and must be converted to °C-days.

$$\begin{aligned} F &= \frac{(24 \text{ h}\cdot\text{d})(7049°\text{F}\cdot\text{d})\left(\frac{5°\text{C}}{9°\text{F}}\right)(23\,500 \text{ W})}{(0.55)(21°\text{C} - (-23°\text{C}))\left(10\,350 \ \frac{\text{W}\cdot\text{h}}{\text{m}^3}\right)} \\ &\quad \times (0.64) \\ &= 5644 \text{ m}^3 \quad (5600 \text{ m}^3) \end{aligned}$$

The answer is A.

Practice Exam 1

PROBLEMS

1. A pump with a mass of 45 kg is supported by four springs, each having a spring constant of 1750 N/m. The motor is constrained to allow only vertical movement. The natural frequency of the pump is most nearly

(A) 6 rad/s
(B) 9 rad/s
(C) 12 rad/s
(D) 15 rad/s

2. The purpose of normalizing SA-516 Grade 70 plate is to produce a

(A) uniform product
(B) homogeneous, rigid product
(C) uniform, fine grain practice microstructure
(D) toughened product

3. A 100 mm gage length is marked on an aluminum rod. The rod is strained so that the gage marks are 109 mm apart. The strain is most nearly

(A) 0.001
(B) 0.01
(C) 0.1
(D) 1.0

4. The percent of sulfur that would be present if it were used to cross-link at every possible point in polyethylene is

(A) 12%
(B) 28%
(C) 46%
(D) 53%

5. A 115 kg motor turns at 1800 rpm, and it is mounted on a pad having a stiffness of 500 kN/m. Due to an unbalanced condition a periodic force of 85 N is applied in a vertical direction, once per revolution. Neglecting damping and horizontal movement, the amplitude of vibration is

(A) 0.024 mm
(B) 0.21 mm
(C) 0.79 mm
(D) 1.3 mm

6. A homogeneous disk of 5 cm radius and 10 kg mass rotates on an axle AB of length 0.5 m and rotates about a fixed point A. The disk is constrained to roll on a horizontal floor.

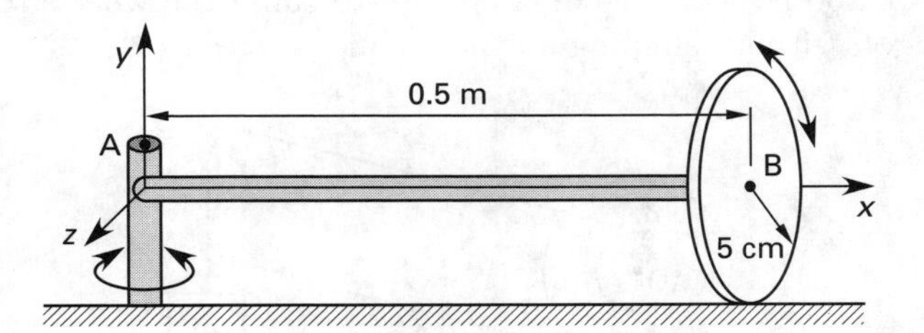

Given an angular velocity of 30 rad/s in the x direction and −3 rad/s in the y direction, the angular momentum of the disk about point A is

(A) 0.375 kg·m^2/s in the x direction, −2.48 kg·m^2/s in the z direction
(B) 0.375 kg·m^2/s in the x direction, −7.52 kg·m^2/s in the y direction
(C) 7.5 kg·m^2/s in the x direction, −9.375 kg·m^2/s in the y direction
(D) 7.5 kg·m^2/s in the x direction, −9.375 kg·m^2/s in the z direction

7. A 2 g bullet with velocity of 100 m/s strikes a 1000 g wooden block moving at a velocity of 10 m/s in the same direction as the bullet. The bullet imbeds itself in the block upon impact. The maximum velocity of the block is most nearly

(A) 5.0 m/s
(B) 10 m/s
(C) 15 m/s
(D) 20 m/s

8. A solid disk 4 m in diameter rolls on a horizontal surface, as shown. At a particular instant, the angular velocity and acceleration of the disk are 10 rad/s (counterclockwise) and 3 rad/s^2 (clockwise), respectively.

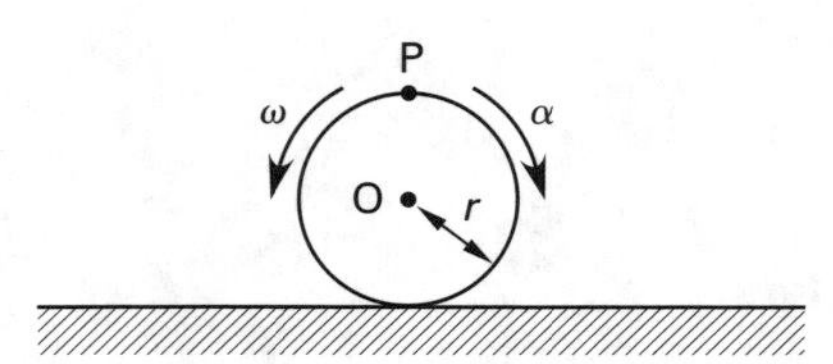

At this instant, the acceleration of point P located at the very top of the disk is most nearly

(A) 100 m/s^2
(B) 150 m/s^2
(C) 200 m/s^2
(D) 250 m/s^2

9. Two splined wheels of radius 20 cm and 50 cm are connected to a mass by a cable, as shown. The cable is roped around the small wheel. The mass exerts a force of 200 N, which is large enough to cause the wheels to roll without slipping on the incline.

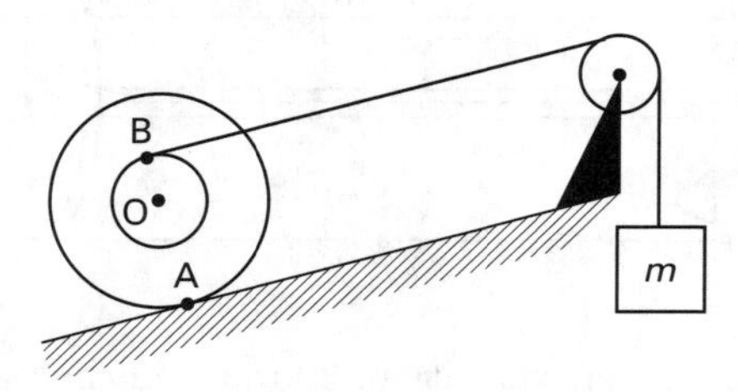

If the mass moves down by 10 cm, the displacement of the wheel is most nearly

(A) 5 cm
(B) 7 cm
(C) 10 cm
(D) 30 cm

10. A uniform disk of 10 kg mass and 0.5 m diameter rolls without slipping on a flat horizontal surface, as shown.

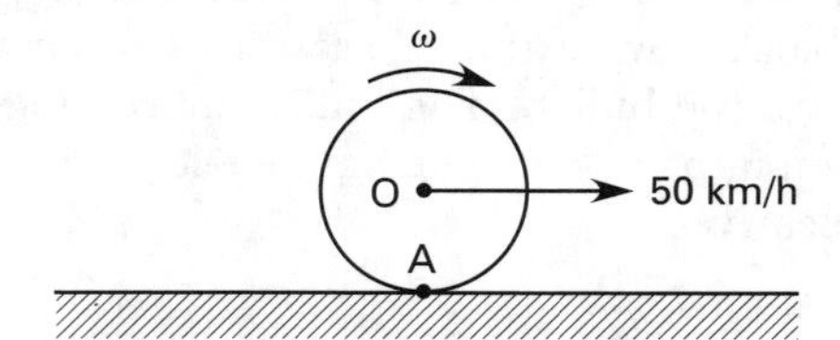

When its horizontal velocity is 50 km/h, the total kinetic energy of the disk is most nearly

(A) 1000 J
(B) 1200 J
(C) 1400 J
(D) 1600 J

11. A two-bar linkage rotates about the pivot point O, as shown. The length of members AB and OA are 2.0 m and 2.5 m, respectively. The angular velocity and acceleration of member OA are $\omega_{OA} = 0.8$ rad/s counterclockwise and $\alpha_{OA} = 0$. The angular velocity of member AB is $\omega_{AB} = 1.2$ rad/s clockwise, and the acceleration of member AB is $\alpha_{AB} = 3$ rad/s^2 counterclockwise.

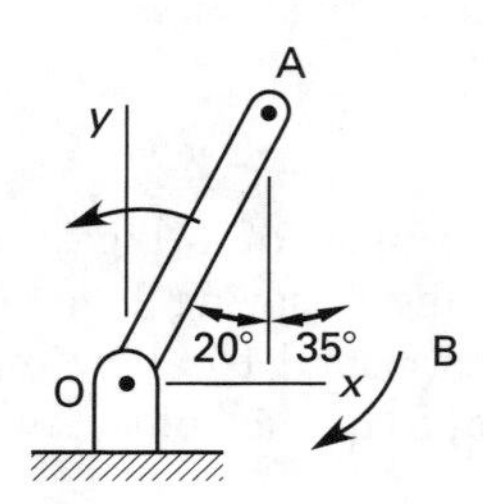

When the bars are in the position shown, the magnitude of the acceleration of point B (i.e., the tip) is most nearly

(A) 3 m/s^2
(B) 5 m/s^2
(C) 8 m/s^2
(D) 10 m/s^2

12. A flywheel with a center hub is used to lower a mass of 150 kg, as shown. The radius of the hub is 50 cm, and the mass moment of inertia of the combined flywheel and hub is 20 kg·m^2. All frictional losses are negligible. At a particular instant, the velocity of the mass is 3 m/s downward.

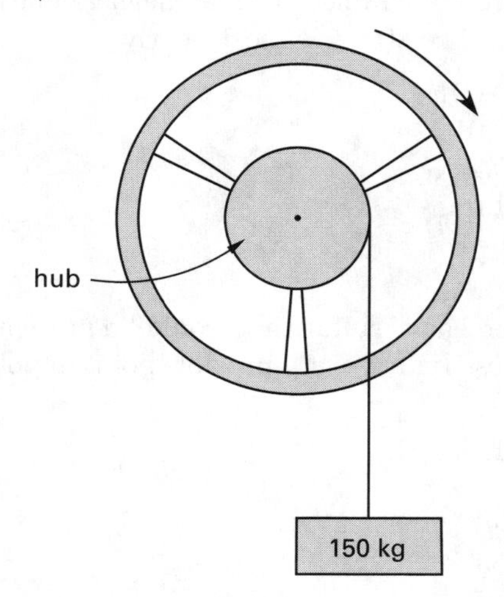

The velocity of the mass after 2 additional meters of travel is most nearly

(A) 4.5 m/s
(B) 6.0 m/s
(C) 9.5 m/s
(D) 14 m/s

13. Superheated steam at a pressure of 1.0 MPa and a temperature of 250°C enters the turbine of an ideal regenerative cycle with one feedwater heater. Steam is extracted at 0.2 MPa to heat the feedwater in an open heater. The rest of the steam expands to the condenser pressure of 10 kPa. Neglecting pump work, the mass of the extracted steam per kilogram of steam entering the turbine is most nearly

(A) 0.09 kg/kg
(B) 0.13 kg/kg
(C) 0.17 kg/kg
(D) 0.20 kg/kg

14. An air-standard diesel cycle has a compression ratio of 18 and a cutoff ratio of 2.2. At the beginning of the isentropic compression process, the temperature of the air is 310K. Assuming the ratio of specific heats, k, is 1.4, the maximum temperature in the cycle is most nearly

(A) 1950K
(B) 2040K
(C) 2170K
(D) 2250K

15. The pressure ratio of an ideal air-standard gas turbine cycle is 7:1. Air enters the compressor at a temperature of 300K and a pressure of 95 kPa. The heat input to the cycle is 700 kJ/kg of air. Assuming constant specific heats, the thermal efficiency is most nearly

(A) 36%
(B) 43%
(C) 48%
(D) 50%

16. A pump is used to move water through a horizontal pipe at a rate of 0.03 m^3/s. The pressure and diameter upstream of the pump are 150 kPa and 100 mm, respectively. The values downstream are 600 kPa and 50 mm, respectively. There are 200 J/kg of frictional losses in the system. The hydraulic efficiency of the pump is most nearly

(A) 70%
(B) 75%
(C) 80%
(D) 85%

17. A fan is used to move air through an air-conditioning duct at a rate of 2.8 m^3/s. The system operating point for the fan, point 1 on the characteristic curve, is indicated. The required power is 1.5 kW. When the speed of the fan is increased to a certain value, it is noticed that the power requirement is tripled.

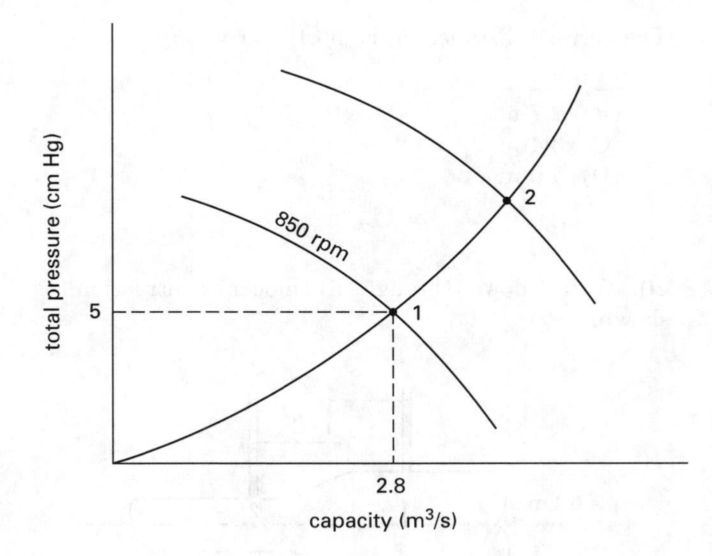

The ratio of the new fan capacity and the total pressure Q_2/p_2 is most nearly

(A) 0.1 m^3/s·cm Hg
(B) 0.2 m^3/s·cm Hg
(C) 0.3 m^3/s·cm Hg
(D) 0.4 m^3/s·cm Hg

18. Air enters a compressor at a pressure of 101 kPa and exits at a pressure of 800 kPa. The air velocity at the entrance is 2 m/s, and the mass flow rate is 5 kg/s. The losses are 10% of the required compressor work, and the increase in kinetic energy across the compressor is 10%. Assume the air density remains constant at 1.23 kg/m^3. If the compressor efficiency is 85%, the power required by the compressor is most nearly

(A) 1.5 MW
(B) 3.0 MW
(C) 3.7 MW
(D) 6.0 MW

19. A vertical jet of water just supports a flat plate having a mass of 1.2 kg, as shown. The nozzle diameter is 2 cm, and the water leaves the nozzle at a speed of 8.5 m/s.

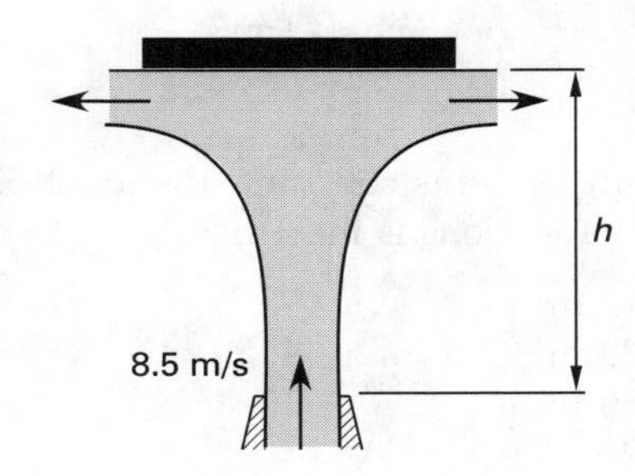

The vertical distance, h, is most nearly

(A) 2.2 m
(B) 2.7 m
(C) 3.4 m
(D) 4.0 m

20. Water flows through a smooth contraction as shown.

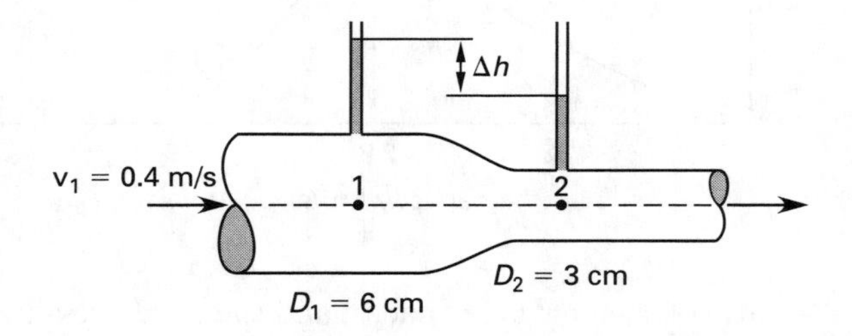

The difference in the fluid heights, Δh, is most nearly

(A) 0.080 m
(B) 0.10 m
(C) 0.12 m
(D) 0.15 m

21. Water flows steadily upward through a diverging tube. At section 1 the diameter is 2 cm, and the velocity is 2.5 m/s. At a subsequent section, the diameter is 8 cm.

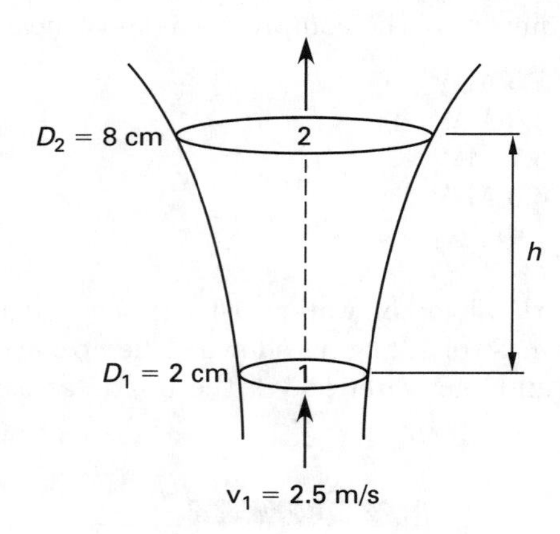

If the pressure remains constant, the axial distance between the two sections is most nearly

(A) 0.12 m
(B) 0.32 m
(C) 0.80 m
(D) 1.2 m

22. A tank having a uniform cross-sectional area of 3 m^2 contains water at 300K with a depth of 2.2 m. The tank is open to the atmosphere. The water is drained through a 0.02 m^2 hole in the bottom of the tank. The time it takes to completely drain the tank is most nearly

(A) 50 s
(B) 75 s
(C) 100 s
(D) 150 s

23. Air at 300K and 101.3 kPa and with a velocity of 22 m/s is flowing over a transverse cylinder 10 cm in diameter. If the kinematic viscosity of the air is 15.89×10^{-6} m^2/s, the drag force per unit length of the cylinder is most nearly

(A) 37 N/m
(B) 43 N/m
(C) 52 N/m
(D) 54 N/m

24. Water at 300K flows through a horizontal 1.5 mm diameter tube. The viscosity of the water is 855×10^{-6} N·s/m^2. What is the maximum pressure drop per meter length of tube such that flow remains laminar?

(A) 9.0 kPa/m
(B) 12 kPa/m
(C) 16 kPa/m
(D) 18 kPa/m

25. The temperature distribution across a 0.3 m thick wall at a certain moment is

$$T = 200 - 200x + 30x^2$$

T is in °C, and x is in meters measured from one side of the wall. The coefficient of thermal conductivity of the wall is 2 W/m·K. The net heat transfer to the wall per unit area is most nearly

(A) 27 W/m^2
(B) 31 W/m^2
(C) 36 W/m^2
(D) 43 W/m^2

26. A stainless steel tube (3 cm inside diameter and 5 cm outside diameter) is covered with 4 cm of insulation. The thermal conductivities of steel and insulation are 20 W/m·K and 0.06 W/m·K, respectively. If the inside wall temperature of the tube is 500°C, and the outside temperature of the insulation is 50°C, what is the heat loss per meter of the tube?

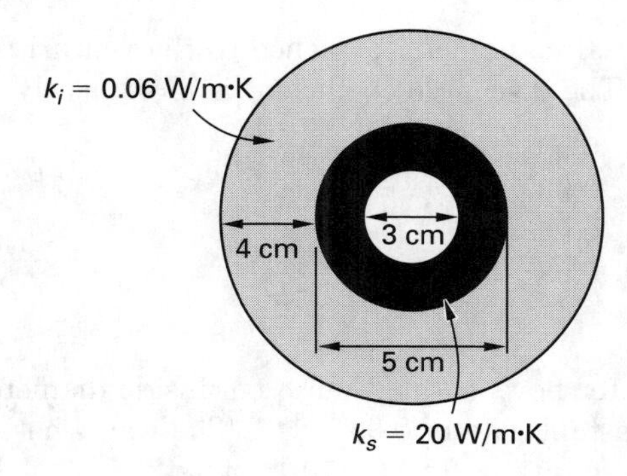

(A) 120 W/m
(B) 140 W/m
(C) 160 W/m
(D) 180 W/m

27. A solid copper sphere 8 cm in diameter is initially at 400°C when it is suddenly exposed to air at 25°C. The average convective heat transfer coefficient is 25 W/m^2. The density, specific heat, and thermal conductivity of copper are 8933 kg/m^3, 410 J/kg·K, and 380 W/m·K, respectively. What is the time required to cool the center of the sphere to 200°C?

(A) 12 min
(B) 16 min
(C) 20 min
(D) 25 min

28. Air at 300K flows at 0.45 m/s over a square flat plate 1 m on each side. The average kinematic viscosity of the air is 20.92×10^{-6} m^2/s, the Prandtl number is 0.7, and the thermal conductivity is 30×10^{-3} W/m·K. If the surface temperature of the plate is 400K, the average heat transfer coefficient is most nearly

(A) 2.5 W/m^2·K
(B) 5.0 W/m^2·K
(C) 7.5 W/m^2·K
(D) 8.2 W/m^2·K

29. A counterflow heat exchanger is used to cool 0.1 kg/s of oil from 100°C to 70°C. Cooling water enters the heat exchanger at 30°C and leaves at 70°C. Assuming the specific heat of oil is 1.9 kJ/kg·K, and the overall heat transfer coefficient is 0.32 kW/m^2·K, the required heat exchanger area is most nearly

(A) 0.1 m^2
(B) 0.3 m^2
(C) 0.5 m^2
(D) 0.7 m^2

30. Two plates of equal area are placed parallel to each other in a vacuum. One plate has an emissivity of 0.2 and temperature of 700K, and the other plate has an emissivity of 0.4 and a temperature of 500K. If the view factor is 0.8, the radiation heat flux will most nearly be

(A) 1.0 kW/m^2
(B) 1.5 kW/m^2
(C) 2.0 kW/m^2
(D) 2.5 kW/m^2

31. In a pearlitic SAE-1080 steel, the cementite platelets are 4×10^{-5} cm thick, and the ferrite platelets are 14×10^{-5} cm thick. The density of ferrite is 7.87 g/cm^3, and the density of cementite is 7.66 g/cm^3. The volume percentage of the cementite in the steel is most nearly

(A) 5.0%
(B) 8.0%
(C) 12%
(D) 18%

32. One-half of an electrochemical cell consists of a pure nickel electrode in a 0.001 molal Ni^{2+} solution, and the other half is a cadmium electrode immersed in a 0.5 molal Cd^{2+} solution. The electrode potential is given by $E = E_0 + 0.0296 \log(c)$, where E_0 is the standard electrode potential and c is the molal concentration of the ions in the solution. The cell potential is most nearly

(A) −0.65 V
(B) −0.15 V
(C) −0.073 V
(D) +0.75 V

33. A 5 mm diameter rod is subjected to a 1.5 kN tensile load. The yield stress, modulus of elasticity, Poisson ratio, and length are 145 MPa, 70 GPa, 0.33, and 10 cm, respectively. The change in diameter of the rod is most nearly

(A) 0.002 mm
(B) 0.004 mm
(C) 0.008 mm
(D) 0.01 mm

34. From the data shown, the pressure difference between tanks A and B is most nearly

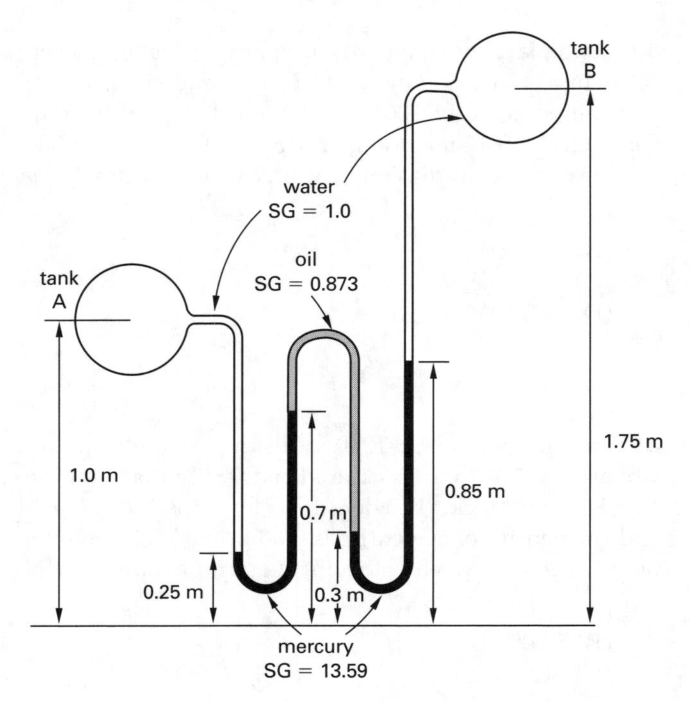

(A) 110 kPa
(B) 120 kPa
(C) 130 kPa
(D) 140 kPa

35. Water flows through a 10 cm inside diameter pipe, as shown.

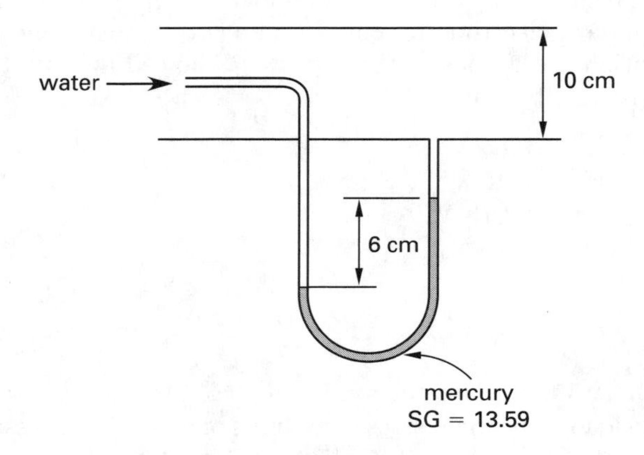

If the deflection of the manometer is 6 cm of mercury, the flow rate is most nearly

(A) 0.03 m^3/s
(B) 0.05 m^3/s
(C) 0.07 m^3/s
(D) 0.09 m^3/s

36. Water flows at a constant rate of 0.0045 m^3/s through a horizontal venturi meter. The diameters at the inlet and throat are 10 cm and 3 cm, respectively. The difference in levels of the mercury columns of a differential manometer attached to the venturi meter is 20 cm. The discharge coefficient is most nearly

(A) 0.90
(B) 0.93
(C) 0.95
(D) 0.98

37. Water flows through an 8 cm inside diameter pipe at a constant rate of 0.03 m^3/s. The water has a kinematic viscosity of 9.609×10^{-5} m^2/s.

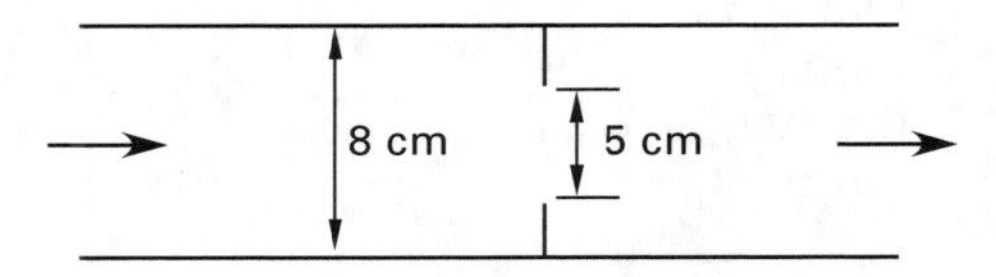

If a 5 cm diameter sharp-edged orifice plate is inserted in the pipe, the static pressure drop across the orifice is most nearly

(A) 200 kPa
(B) 260 kPa
(C) 310 kPa
(D) 380 kPa

38. Oil with a kinematic viscosity of 1120×10^{-6} m^2/s flows at a velocity of 12.5 m/s in a 30 cm diameter pipe. To achieve dynamic similarity, a test is run using water at 17°C in a 2.5 cm diameter pipe. The viscosity of 17°C water is 1.08×10^{-6} m^2/s. The velocity of the water should be most nearly

(A) 0.15 m/s
(B) 0.50 m/s
(C) 1.0 m/s
(D) 1.5 m/s

39. Air at a temperature of 320K flows at a supersonic speed in a wind tunnel. If the Mach number is 1.5, the velocity of air is most nearly

(A) 400 m/s
(B) 450 m/s
(C) 480 m/s
(D) 540 m/s

40. An aluminum tube is capped at two ends by two rigid plates. The plates are held in place by a steel bolt and a nut, as shown. The pitch of the bolt is 2 mm, and its cross-sectional area is 2 cm^2. The cross-sectional area of the tube wall is 4 cm^2. After being snugly fit, the nut is turned a quarter of a turn, compressing the tube.

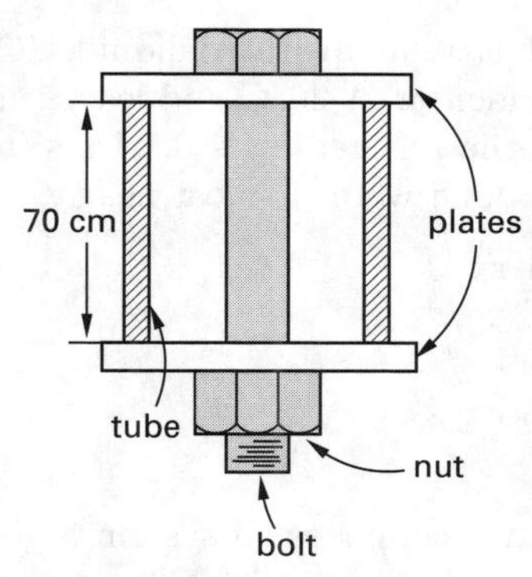

The stress in the bolt is most nearly

(A) 40 MPa
(B) 50 MPa
(C) 60 MPa
(D) 70 MPa

41. A bar of circular cross section is bent at both ends and loaded at the free ends, as shown. The length and the diameter of the bar are 2 m and 6 cm, respectively, and the allowable tensile stress is 1.2 MPa.

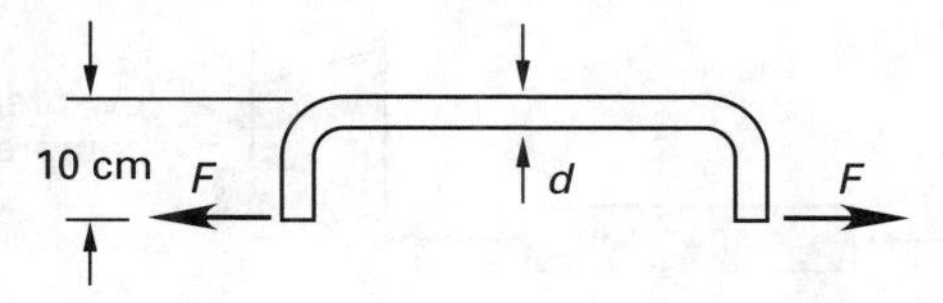

The maximum load that can safely be applied at the free end is most nearly

(A) 240 N
(B) 280 N
(C) 330 N
(D) 400 N

42. A pivoted shaft with a solid circular cross section is suspended from a rigid surface, as shown. The temperature of the shaft is increased 500°C, causing it to elongate. The thermal coefficient of expansion and modulus of elasticity are 15×10^{-6} 1/°C and 150 GPa, respectively.

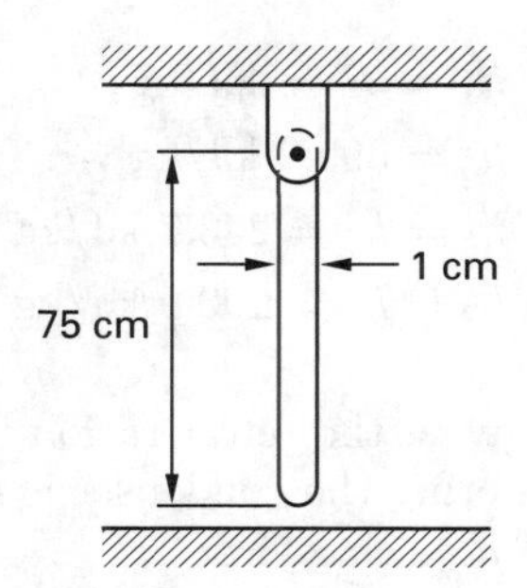

To avoid buckling of the shaft, the clearance between the free end of the shaft and the bottom surface should be most nearly

(A) 5.35 mm
(B) 5.45 mm
(C) 5.55 mm
(D) 5.60 mm

43. A rectangular steel plate is bolted to a rectangular column with four bolts, as shown. A load of 20 kN is applied at the edge, as shown.

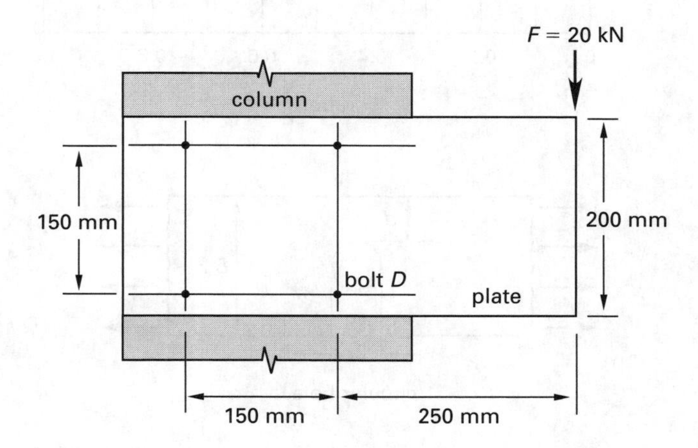

The resulting shear load on bolt D is most nearly

(A) 11 kN
(B) 15 kN
(C) 19 kN
(D) 20 kN

44. A hollow shaft is used to transmit 35 kW of power. The rotational speed is 1000 rpm. The inside and outside diameters are 30 mm and 40 mm, respectively. The torsional shear stress developed in the shaft is most nearly

(A) 40 MPa
(B) 80 MPa
(C) 110 MPa
(D) 130 MPa

45. The plate shown has a concentric circular hole drilled through it. The plate is made from a brittle material whose ultimate stress is 1.5 MPa.

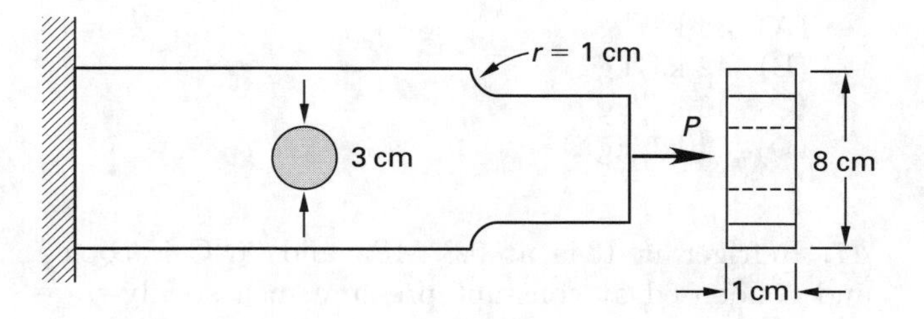

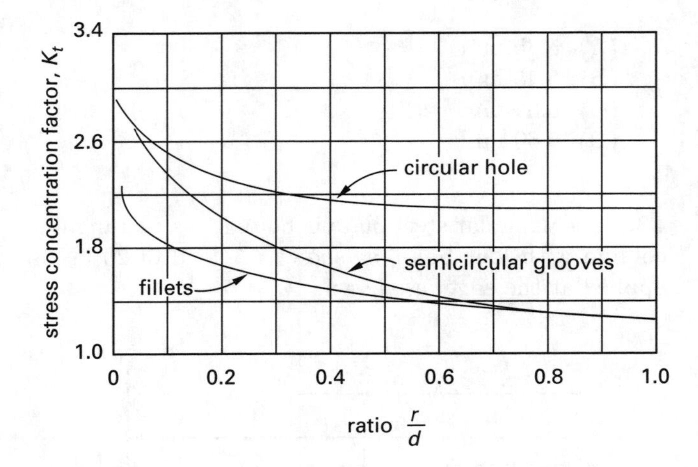

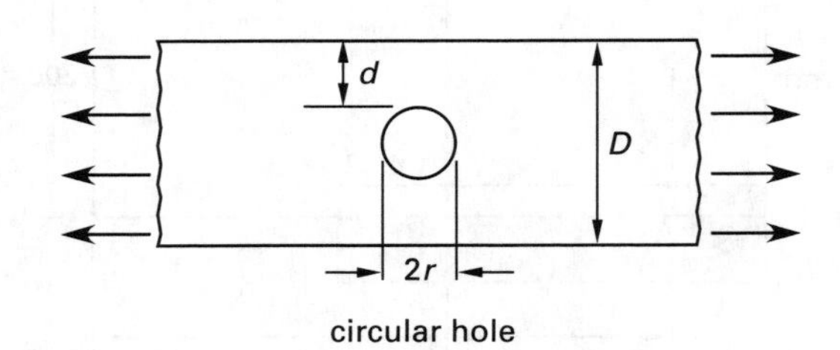

circular hole

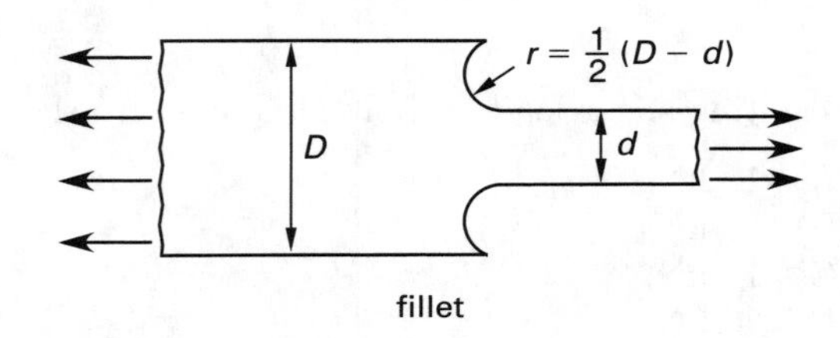

fillet

Using a factor of safety of 2.5 and the stress concentration diagram shown, the allowable tensile load is most nearly

(A) 120 N
(B) 140 N
(C) 150 N
(D) 170 N

46. A vapor-compression refrigeration cycle utilizes R-134a as a working fluid. The refrigerant enters the compressor as a saturated vapor at −10°C and leaves the condenser as a saturated liquid at 40°C. If the compressor efficiency is 80%, the work input to the compressor is most nearly

(A) 36 kJ/kg
(B) 42 kJ/kg
(C) 46 kJ/kg
(D) 52 kJ/kg

47. Refrigerant-134a at 0.8 MPa and 70°C is cooled and condensed at constant pressure in a steady-state process until it is a saturated liquid. Cooling water enters the condenser at 20°C and leaves at 30°C. If the mass flow of the refrigerant is 0.1 kg/s, the mass flow rate of the cooling water is most nearly

(A) 0.51 kg/s
(B) 0.65 kg/s
(C) 0.70 kg/s
(D) 0.75 kg/s

48. A two-stage compression system together with the corresponding p-h diagram is shown.

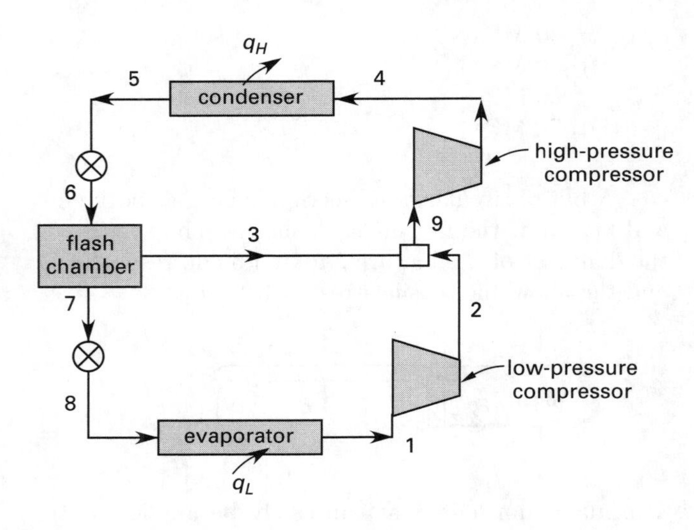

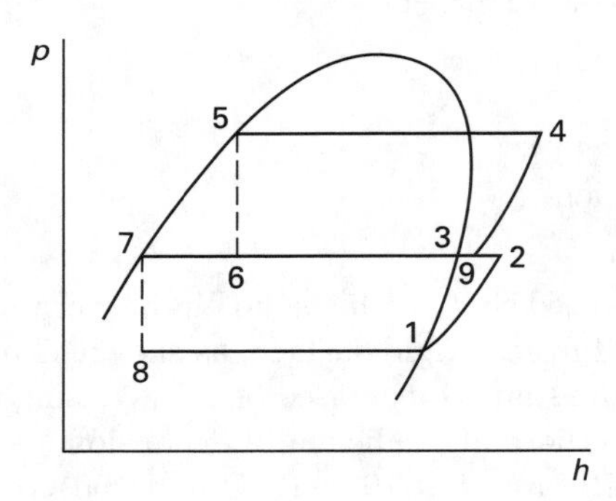

$$h_1 = 388.5\ \text{kJ/kg}$$
$$h_3 = 399.2\ \text{kJ/kg}$$
$$h_5 = h_6 = 243.7\ \text{kJ/kg}$$
$$h_7 = h_8 = 200.9\ \text{kJ/kg}$$

From the data given, the refrigeration load q_L per unit mass of fluid entering the condenser is most nearly

(A) 120 kJ/kg
(B) 135 kJ/kg
(C) 145 kJ/kg
(D) 160 kJ/kg

49. Member A in the truss shown is loaded axially at 30 kN. The allowable shear stress in member B is 600 kPa. Both members of the truss have the same width.

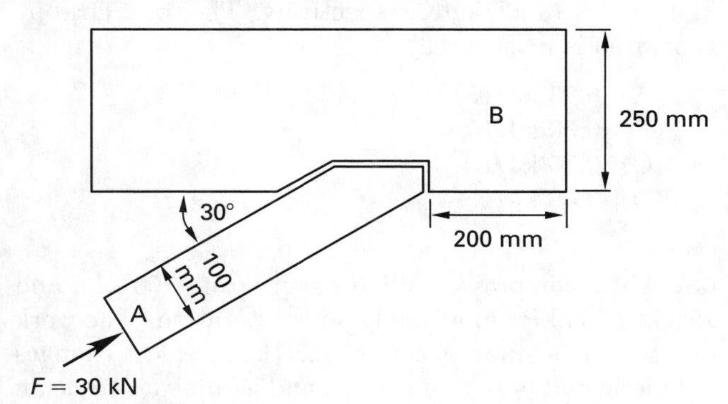

The width of member B to prevent shearing is most nearly

(A) 140 mm
(B) 170 mm
(C) 200 mm
(D) 220 mm

50. The truss shown is loaded at joint B with a vertical load of 90 kN. Both members of the truss have a cross-sectional area of 4 cm^2 and a length of 1 m, and both are made of steel.

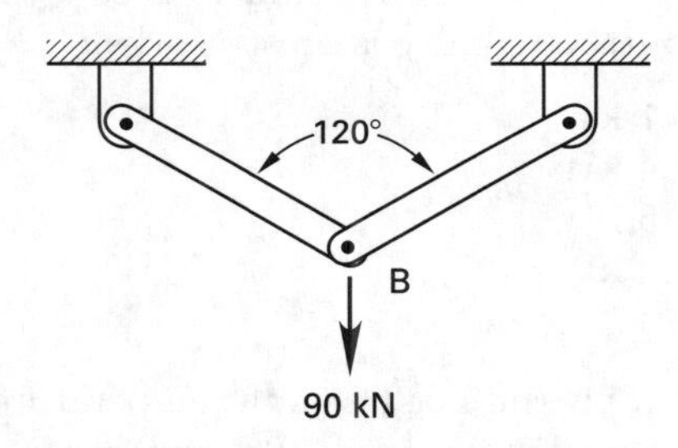

The vertical displacement of joint B is most nearly

(A) 1.0 mm
(B) 2.1 mm
(C) 3.3 mm
(D) 8.6 mm

51. An aluminum bar 60 cm long is placed between two supports that are 60 cm apart. The supports are rigid and fixed and do not exert any initial stress on the bar. The bar is heated such that its temperature changes from 25°C to 100°C. The coefficient of thermal expansion for the aluminum is 23×10^{-6} 1/C°.

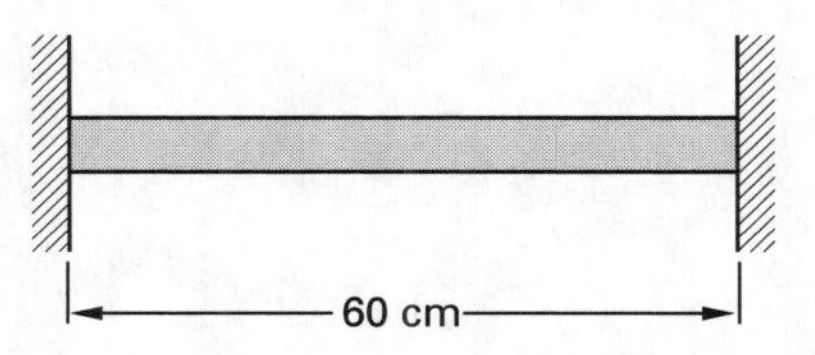

The stress developed at the supports is most nearly

(A) 70 MPa
(B) 120 MPa
(C) 170 MPa
(D) 250 MPa

52. A stepped solid circular shaft is subjected to torques T_1 and T_2, as shown. The shaft diameters (d_1 and d_2) are 80 mm and 60 mm, and the corresponding sectional lengths are 0.6 m and 0.4 m. The material is steel with a shear modulus of elasticity of 80 GPa.

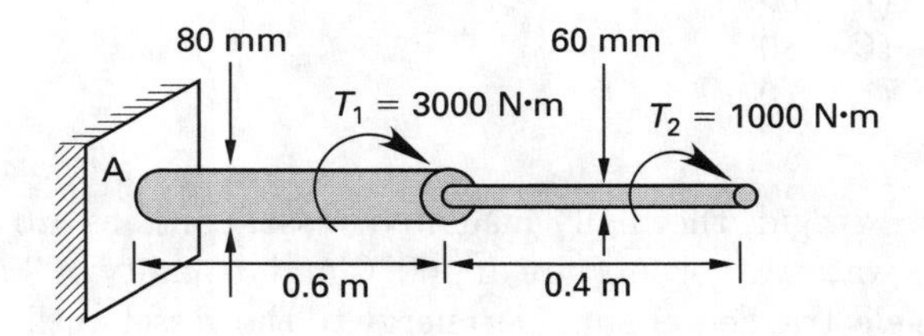

The angle of twist at the free end is most nearly

(A) 0.1°
(B) 0.2°
(C) 0.4°
(D) 0.7°

53. The cylindrical pressure vessel has an inside diameter of 2.5 m and a wall thickness of 15 mm. It is made of steel plates that are welded along a seam that makes an angle of 45° with the longitudinal axis. The internal pressure of the vessel is 8 MPa.

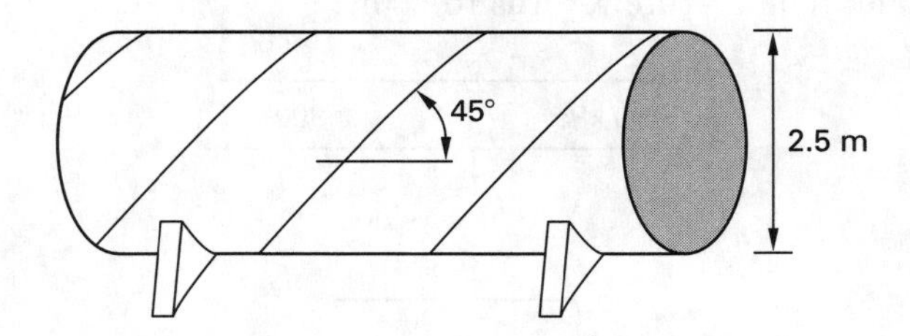

The normal component of the stress along the seam is most nearly

(A) 500 MPa
(B) 600 MPa
(C) 700 MPa
(D) 800 MPa

54. A cantilever beam with a rectangular cross section is subjected to a 150 kN load, as shown. The height and the width of the beam are 150 mm and 75 mm, respectively.

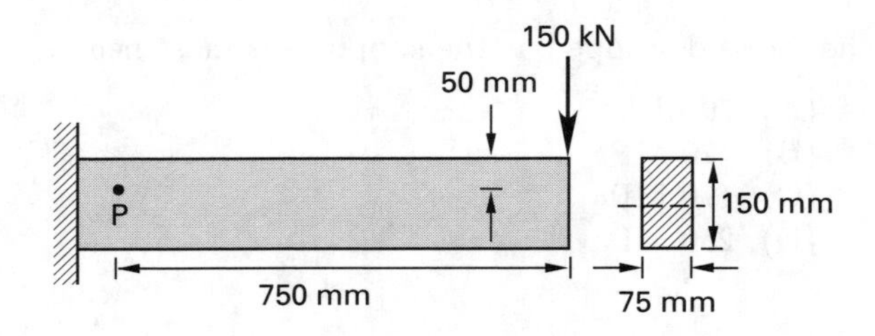

At the point P (50 mm from the top surface), the ratio of the magnitudes of the principal stresses (σ_1/σ_2) is most nearly

(A) 40
(B) 60
(C) 80
(D) 100

55. A rigid, thermally insulated vessel contains 1.0 kg of a water-vapor mixture at 100°C and a quality of 30%. An electric heater supplies energy to the vessel until the mixture becomes saturated vapor. The electric energy supplied is most nearly

(A) 1.0 MJ
(B) 1.2 MJ
(C) 1.5 MJ
(D) 1.6 MJ

56. A tank is connected to a supply manifold through a valve. The tank has a volume of 1 m^3, and it initially contains air at 150 kPa and 300K. The air in the supply manifold has a pressure of 750 kPa and a temperature of 300K. The tank is uninsulated. The valve is opened and left open until the tank pressure stabilizes and the tank temperature returns to 300K.

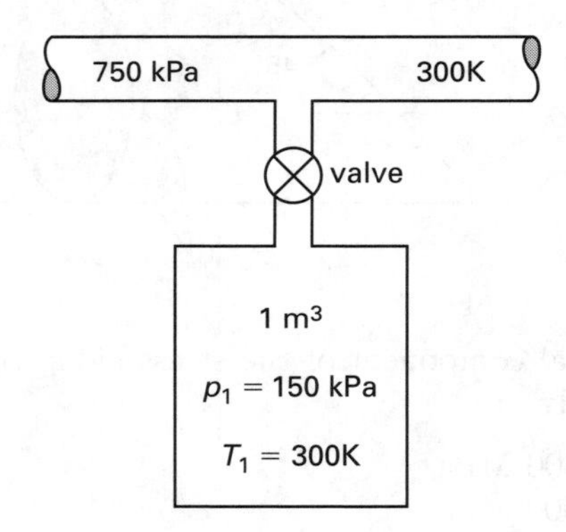

The heat transfer to the surroundings during the process is most nearly

(A) 100 kJ
(B) 150 kJ
(C) 450 kJ
(D) 600 kJ

57. A 5 kg block of copper initially at 200°C is placed in an insulated tank containing 10 kg of water at 25°C. The specific heats of copper and of water are 0.39 kJ/kg·K and 4.18 kJ/kg·K respectively. The total entropy generated is most nearly

(A) 0.23 kJ/K
(B) 0.30 kJ/K
(C) 0.35 kJ/K
(D) 0.38 kJ/K

58. Air is compressed adiabatically from 100 kPa and 25°C to 300 kPa in a steady-state operation. The work input to the compressor is 150 kJ/kg. Neglect changes in kinetic and potential energy, and assume air to be an ideal gas with constant specific heats. The net change in entropy per unit mass for the process is most nearly

(A) 0.02 kJ/kg·K
(B) 0.06 kJ/kg·K
(C) 0.09 kJ/kg·K
(D) 2 kJ/kg·K

59. A steady flow of air enters an adiabatic nozzle with negligible velocity at a pressure of 180 kPa and a temperature of 65°C. The mass flow rate is 1.0 kg/s, the exit pressure is 100 kPa, and the exit velocity is 300 m/s. The temperature of the environment is 10°C. The power lost due to irreversibility is most nearly

(A) 6.7 kW
(B) 7.4 kW
(C) 8.5 kW
(D) 9.2 kW

60. Air at a pressure of 1 atm, a temperature of 45°C, and a relative humidity of 10% enters an adiabatic humidifier operating at steady state. If the dry- and wet-bulb temperatures at the exit are 33.5°C and 31°C, respectively, the increase in the specific humidity is most nearly

(A) 0.015 kg/kg
(B) 0.019 kg/kg
(C) 0.021 kg/kg
(D) 0.022 kg/kg

SOLUTIONS

1. Calculate the total spring constant.

$$k = (4)\left(1750\ \frac{\text{N}}{\text{m}}\right) = 7000\ \text{N/m}$$

Calculate the static deflection.

$$\delta_{\text{st}} = \frac{W}{k} = \frac{mg}{k} = \frac{(45\ \text{kg})\left(9.81\ \frac{\text{m}}{\text{s}^2}\right)}{7000\ \frac{\text{N}}{\text{m}}} = 0.063\ \text{m}$$

Calculate the natural frequency.

$$\omega = \sqrt{\frac{g}{\delta_{\text{st}}}} = \sqrt{\frac{9.81\ \frac{\text{m}}{\text{s}^2}}{0.063\ \text{m}}} = 12.48\ \text{rad/s} \quad (12\ \text{rad/s})$$

The answer is C.

2. The purpose of normalizing is to produce a uniform, fine grain practice microstructure. Normalizing is commonly done to prevent hydrogen stress cracking.

The answer is C.

3.

$$\varepsilon = \frac{\Delta L}{L_o} = \frac{109\ \text{mm} - 100\ \text{mm}}{100\ \text{mm}} = 0.09 \quad (0.1)$$

The answer is C.

4. The formula for polyethylene is C_2H_4. The molecular weight is

$$(2)(12) + (4)(1) = 28$$

There are 28 grams of polyethylene in one mole.

One mole of sulfur is 32 grams.

Therefore, the percent of sulfur present would be

$$\frac{32}{32+28} = 0.53 \times 100\% = 53\%$$

The answer is D.

5. The natural frequency is

$$\omega_n = \sqrt{\frac{k}{m}} = \sqrt{\frac{\left(500\ \frac{\text{kN}}{\text{m}}\right)\left(1000\ \frac{\text{N}}{\text{kN}}\right)}{115\ \text{kg}}} = 66\ \text{rad/s}$$

The forcing frequency is

$$\omega_f = \left(\frac{1800\ \frac{\text{rev}}{\text{min}}}{60\ \frac{\text{s}}{\text{min}}}\right)\left(2\pi\ \frac{\text{rad}}{\text{rev}}\right) = 188.5\ \text{rad/s}$$

The pseudo-static deflection is

$$\delta_{\text{pst}} = \frac{F_o}{k} = \frac{85\ \text{N}}{\left(500\ \frac{\text{kN}}{\text{m}}\right)\left(1000\ \frac{\text{N}}{\text{kN}}\right)} = 1.7 \times 10^{-4}\ \text{m}$$

The magnification factor is

$$\beta = \left|\frac{1}{1-\left(\frac{\omega_f}{\omega_n}\right)^2}\right| = \left|\frac{1}{1-\left(\frac{188.5\ \frac{\text{rad}}{\text{s}}}{66\ \frac{\text{rad}}{\text{s}}}\right)^2}\right| = 0.14$$

The amplitude of vibration is then

$$D = \beta\delta_{\text{pst}} = (0.14)\left(1.7 \times 10^{-4}\ \text{m}\right)\left(1000\ \frac{\text{mm}}{\text{m}}\right) = 0.0238\ \text{mm} \quad (0.024\ \text{mm})$$

The answer is A.

6. Assume the axle to be part of the disk; then the disk has a fixed point at A. Since the x, y, and z axes are principle axes of inertia for the disk, the momentum in the x direction is

$$h_x = I_x\omega_x = \tfrac{1}{2}mr^2\omega_x = \left(\frac{1}{2}\right)(10\ \text{kg})(0.05\ \text{m})^2\left(30\ \frac{\text{rad}}{\text{s}}\right) = 0.375\ \text{kg}\cdot\text{m}^2/\text{s}$$

The momentum in the y direction is

$$\begin{aligned} h_y &= I_y\omega_y \\ &= \left(mL^2 + \tfrac{1}{4}mr^2\right)\omega_y \\ &= \left((10\text{ kg})(0.5\text{ m})^2 + \left(\frac{1}{4}\right)(10\text{ kg})(0.05\text{ m})^2\right) \\ &\quad \times \left(-3\ \frac{\text{rad}}{\text{s}}\right) \\ &= -7.5188\ \text{kg·m}^2/\text{s} \quad (-7.52\ \text{kg·m}^2/\text{s}) \end{aligned}$$

The momentum in the z direction is

$$\begin{aligned} h_z &= I_z\omega_z = \left(mL^2 + \tfrac{1}{4}mr^2\right)(0) \\ &= 0 \end{aligned}$$

Therefore, $h_A = 0.375\ \text{kg·m}^2/\text{s}$ in the x direction, $-7.52\ \text{kg·m}^2/\text{s}$ in the y direction.

The answer is B.

7. Applying the law of conservation of momentum,

$$\begin{aligned} m_{\text{block}}v_{\text{block}} + m_{\text{bullet}}v_{\text{bullet}} &= (m_{\text{block}} + m_{\text{bullet}})v_{\text{max}} \\ v_{\text{max}} &= \frac{m_{\text{block}}v_{\text{block}} + m_{\text{bullet}}v_{\text{bullet}}}{m_{\text{block}} + m_{\text{bullet}}} \\ &= \frac{(1000\text{ g})\left(10\ \frac{\text{m}}{\text{s}}\right) + (2\text{ g})\left(100\ \frac{\text{m}}{\text{s}}\right)}{1000\text{ g} + 2\text{ g}} \\ &= 10.18\ \text{m/s} \quad (10\ \text{m/s}) \end{aligned}$$

The answer is B.

8.

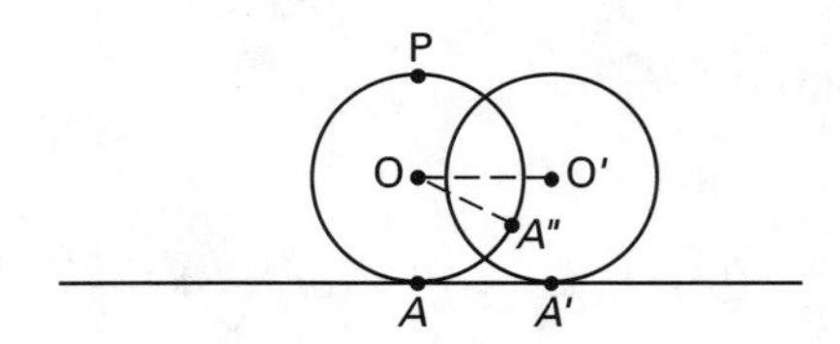

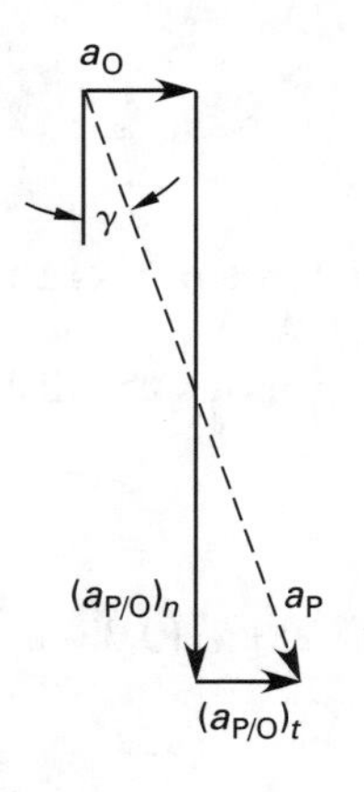

$$AA'' = r\theta$$

Let $AA' = AA'' = S = r\theta$. The linear displacement of the center O is

$$\begin{aligned} d &= \text{OO}' \\ v_O &= r\omega = (2\text{ m})\left(10\ \frac{\text{rad}}{\text{s}}\right) = 20\ \text{m/s} \\ a_O &= r\alpha = (2\text{ m})\left(3\ \frac{\text{rad}}{\text{s}^2}\right) \\ &= 6\ \text{m/s}^2 \quad \text{[to the right]} \end{aligned}$$

$$\begin{aligned} v_P &= v_{P/O} + v_O \\ v_{P/O} &= (\text{OP})\omega = (2\text{ m})\left(10\ \frac{\text{rad}}{\text{s}}\right) = 20\ \text{m/s} \\ v_P &= 20\ \frac{\text{m}}{\text{s}} + 20\ \frac{\text{m}}{\text{s}} = 40\ \text{m/s} \quad \text{[to the left]} \\ a_P &= \left(a_{P/O}\right)_t + \left(a_{P/O}\right)_n + a_O \\ \left(a_{P/O}\right)_t &= (\text{OP})\alpha = (2\text{ m})\left(3\ \frac{\text{rad}}{\text{s}^2}\right) \\ &= 6\ \text{m/s}^2 \quad \text{[to the right]} \\ \left(a_{P/O}\right)_n &= (\text{OP})\omega^2 = (2\text{ m})\left(10\ \frac{\text{rad}}{\text{s}}\right)^2 \\ &= 200\ \text{m/s}^2 \quad \text{[downward]} \\ a_P &= \sqrt{\left(a_O + (a_{P/O})_t\right)^2 + (a_{P/O})_n^2} \\ &= \sqrt{\left(6\ \frac{\text{m}}{\text{s}^2} + 6\ \frac{\text{m}}{\text{s}^2}\right)^2 + \left(200\ \frac{\text{m}}{\text{s}^2}\right)^2} \\ &= 200.4\ \text{m/s}^2 \quad (200\ \text{m/s}^2) \end{aligned}$$

The answer is C.

9. The angle of rotation is the same for both splined wheels.

The circumferential fractions are the same.

$$\begin{aligned} \frac{d}{C} &= \frac{10\text{ cm}}{2\pi(20\text{ cm})} = \frac{x}{2\pi(50\text{ cm})} \\ x &= 25\text{ cm} \quad (30\text{ cm}) \end{aligned}$$

The answer is D.

10.

$$\begin{aligned} v_O &= \left(50\ \frac{\text{km}}{\text{h}}\right)\left(1000\ \frac{\text{m}}{\text{km}}\right)\left(\frac{1\text{ h}}{3600\text{ s}}\right) \\ &= 13.89\ \text{m/s} \end{aligned}$$

$$\begin{aligned} \text{angular speed} = \omega &= \frac{v_O}{r} = \frac{13.89\ \frac{\text{m}}{\text{s}}}{\frac{0.5\text{ m}}{2}} \\ &= 55.56\ \text{rad/s} \end{aligned}$$

$$\begin{aligned}
\text{KE}_{\text{total}} &= \text{KE}_{\text{translation}} + \text{KE}_{\text{rotation}} \\
&= \tfrac{1}{2}m\text{v}_0^2 + \tfrac{1}{2}I_0\omega^2 \\
&= \tfrac{1}{2}m\text{v}_0^2 + \tfrac{1}{2}\left(\tfrac{1}{2}mR^2\right)\omega^2 \\
&= \left(\frac{1}{2}\right)(10\text{ kg})\left(13.89\ \frac{\text{m}}{\text{s}}\right)^2 \\
&\quad + \left(\frac{1}{2}\right)\left(\left(\frac{1}{2}\right)(10\,\text{kg})\left(\frac{0.5\,\text{m}}{2}\right)^2\right) \\
&\quad \times \left(55.56\ \frac{\text{rad}}{\text{s}}\right)^2 \\
&= 1447\,\text{J} \quad (1400\,\text{J})
\end{aligned}$$

The answer is C.

11.

$$\begin{aligned}
\mathbf{a}_{\text{A}} &= \boldsymbol{\omega}_{\text{OA}} \times (\boldsymbol{\omega}_{\text{OA}} \times \mathbf{r}_{\text{A/O}}) + (\boldsymbol{\alpha}_{\text{OA}} \times \mathbf{r}_{\text{A/O}}) \\
&= \left(0.8\mathbf{k}\ \frac{\text{rad}}{\text{s}}\right) \\
&\quad \times \left(\begin{array}{c}\left(0.8\mathbf{k}\ \frac{\text{rad}}{\text{s}}\right)(2.5\text{ m}) \\ \times\ (\mathbf{i}\sin 20^\circ + \mathbf{j}\cos 20^\circ)\end{array}\right) + 0 \\
&= -0.547\mathbf{i} - 1.504\mathbf{j} \quad [\text{toward O}] \\
\mathbf{a}_{\text{B/A}} &= \boldsymbol{\omega}_{\text{AB}} \times (\boldsymbol{\omega}_{\text{AB}} \times \mathbf{r}_{\text{B/A}}) + (\boldsymbol{\alpha}_{\text{AB}} \times \mathbf{r}_{\text{B/A}}) \\
&= \left(-1.2\mathbf{k}\ \frac{\text{rad}}{\text{s}}\right) \\
&\quad \times \left(\begin{array}{c}\left(-1.2\mathbf{k}\ \frac{\text{rad}}{\text{s}}\right) \\ \times\ (2.0\text{ m})(\mathbf{i}\sin 35^\circ - \mathbf{j}\cos 35^\circ)\end{array}\right) \\
&\quad + \left(3\mathbf{k}\ \frac{\text{rad}}{\text{s}^2}\right) \\
&\quad \times (2.0\text{ m})(\mathbf{i}\sin 35^\circ - \mathbf{j}\cos 35^\circ) \\
&= -1.6519\mathbf{i} + 2.3592\mathbf{j} + 3.4415\mathbf{j} + 4.591\mathbf{i} \\
&= 3.263\mathbf{i} + 5.800\mathbf{j}
\end{aligned}$$

The acceleration of point B with respect to point O is

$$\begin{aligned}
\mathbf{a}_{\text{B}} &= \mathbf{a}_{\text{A}} + \mathbf{a}_{\text{B/A}} \\
&= -0.547\mathbf{i} - 1.504\mathbf{j} + 3.263\mathbf{i} + 5.800\mathbf{j} \\
&= 2.716\mathbf{i} + 4.296\mathbf{j} \\
|\mathbf{a}_{\text{B}}| &= \sqrt{\left(2.716\ \frac{\text{m}}{\text{s}^2}\right)^2 + \left(4.296\ \frac{\text{m}}{\text{s}^2}\right)^2} \\
&= 5.083\text{ m/s}^2 \quad (5\text{ m/s}^2)
\end{aligned}$$

The answer is B.

12. The angular velocity is

$$\begin{aligned}
\omega_1 &= \frac{\text{v}_1}{r} = \frac{3\ \frac{\text{m}}{\text{s}}}{0.5\text{ m}} = 6\text{ rad/s} \\
E_1 &= \tfrac{1}{2}m\text{v}_1^2 + \tfrac{1}{2}I\omega_1^2 \\
&= \left(\frac{1}{2}\right)(150\text{ kg})\left(3\ \frac{\text{m}}{\text{s}}\right)^2 \\
&\quad + \left(\frac{1}{2}\right)(20\text{ kg·m}^2)\left(6\ \frac{\text{rad}}{\text{s}}\right)^2 \\
&= 1035\text{ J} \\
E_2 &= \tfrac{1}{2}m\text{v}_2^2 + \tfrac{1}{2}I\omega_2^2 \\
&= \left(\frac{1}{2}\right)(150\text{ kg})\text{v}_2^2 + \left(\frac{1}{2}\right)(20\text{ kg·m}^2)\left(\frac{\text{v}_2}{0.5\text{ m}}\right)^2 \\
&= (115\text{ kg})\text{v}_2^2
\end{aligned}$$

From the principle of work and energy,

$$\begin{aligned}
E_1 + W_{1-2} &= E_2 \\
W_{1-2} &= F\Delta y = (150\text{ kg})\left(9.81\ \frac{\text{m}}{\text{s}^2}\right)(2\text{ m}) \\
&= 2943\text{ J} \\
1035\text{ J} + 2943\text{ J} &= (115\text{ kg})\text{v}_2^2 \\
\text{v}_2 &= 5.88\text{ m/s} \quad (6.0\text{ m/s})
\end{aligned}$$

The answer is B.

13.

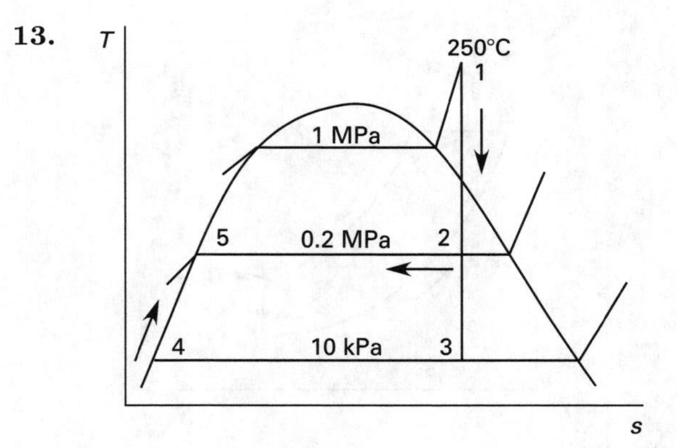

$$\begin{aligned}
p_1 &= 1\text{ MPa} \\
T_1 &= 250^\circ\text{C} \\
h_1 &= 2942.6\text{ kJ/kg} \\
s_1 &= 6.9247\text{ kJ/kg·K} \\
p_2 &= 0.2\text{ MPa} \\
T_2 &\approx 120^\circ\text{C} \quad \left[\begin{array}{l}\text{use properties of saturated} \\ 120^\circ\text{C water for convenience}\end{array}\right] \\
s_2 &= s_1 = s_f + x_2 s_{fg}
\end{aligned}$$

$$x_2 = \frac{s_2 - s_f}{s_{fg}} = \frac{6.9247\ \frac{\text{kJ}}{\text{kg·K}} - 1.5276\ \frac{\text{kJ}}{\text{kg·K}}}{5.6020\ \frac{\text{kJ}}{\text{kg·K}}} = 0.963$$

$$\begin{aligned} h_2 &= h_f + x_2 h_{fg} \\ &= 503.71\ \frac{\text{kJ}}{\text{kg}} + (0.963)\left(2202.6\ \frac{\text{kJ}}{\text{kg}}\right) \\ &= 2625\ \text{kJ/kg} \end{aligned}$$

$$p_4 = 10\ \text{kPa} \quad \left[\begin{array}{l}\text{use properties of saturated} \\ 45^\circ\text{C water for convenience}\end{array}\right]$$

$$h_4 = 188.45\ \text{kJ/kg}$$

$$p_5 = 0.2\ \text{MPa}$$

$$\begin{aligned} h_5 &= h_f \\ &= 503.71\ \text{kJ/kg} \end{aligned}$$

The energy balance equation for the heater is

$$mh_2 + (1 - m)h_4 = h_5$$

$$m = \frac{h_5 - h_4}{h_2 - h_4} = \frac{503.71\ \frac{\text{kJ}}{\text{kg}} - 188.45\ \frac{\text{kJ}}{\text{kg}}}{2625\ \frac{\text{kJ}}{\text{kg}} - 188.45\ \frac{\text{kJ}}{\text{kg}}} = 0.129\ \text{kg/kg} \quad (0.13\ \text{kg/kg})$$

The answer is B.

14.

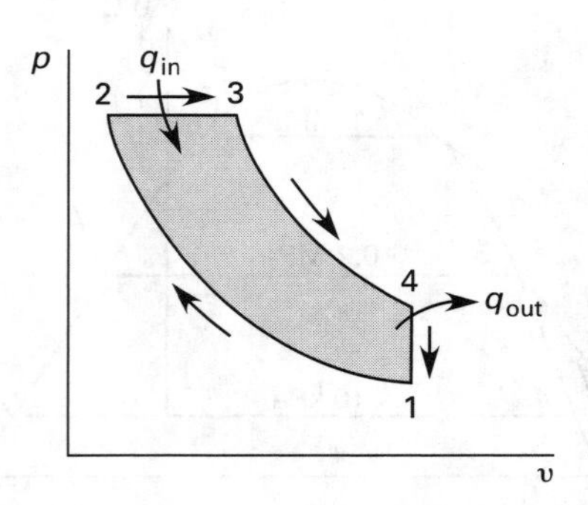

For the compression process,

$$\frac{T_2}{T_1} = \left(\frac{v_1}{v_2}\right)^{k-1}$$

The temperature at the end of the isentropic compression is

$$\begin{aligned} T_2 &= T_1 \left(\frac{v_1}{v_2}\right)^{k-1} = T_1 (r_v)^{k-1} \\ &= (310\text{K})(18)^{1.4-1} \\ &= 985\text{K} \end{aligned}$$

The temperature at the end of the combustion process is

$$\begin{aligned} T_3 &= T_2 \left(\frac{v_3}{v_2}\right) = T_2 r_c \\ &= (985\text{K})(2.2) \\ &= 2167\text{K} \quad (2170\text{K}) \end{aligned}$$

The answer is C.

15.

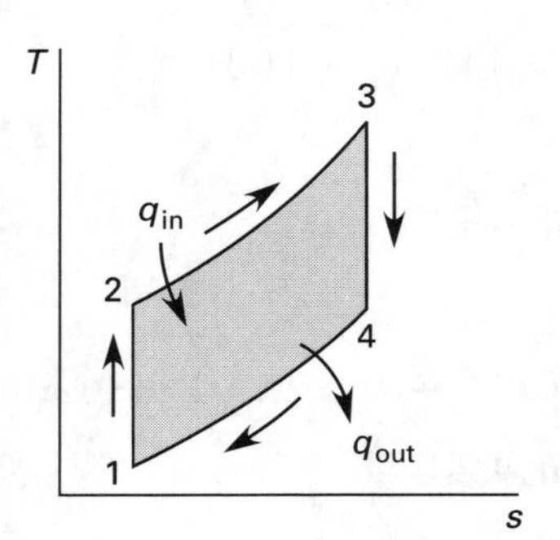

$$\frac{T_2}{T_1} = \left(\frac{p_2}{p_1}\right)^{\frac{k-1}{k}}$$

$$\begin{aligned} T_2 &= (300\text{K})(7)^{\frac{1.4-1}{1.4}} \\ &= 523.1\text{K} \end{aligned}$$

$$q_{2-3} = c_p(T_3 - T_2)$$

$$\begin{aligned} T_3 &= \frac{q_{2-3}}{c_p} + T_2 \\ &= \frac{700\ \frac{\text{kJ}}{\text{kg}}}{1.0035\ \frac{\text{kJ}}{\text{kg·K}}} + 523.1\text{K} \\ &= 1220.7\text{K} \end{aligned}$$

$$\frac{T_4}{T_3} = \left(\frac{p_4}{p_3}\right)^{\frac{k-1}{k}}$$

$$\begin{aligned} T_4 &= (1220.7\text{K})\left(\frac{1}{7}\right)^{\frac{1.4-1}{1.4}} \\ &= 700.1\text{K} \end{aligned}$$

$$\begin{aligned} \eta &= \frac{|W_{\text{actual}}|}{q_{\text{in}}} = \frac{c_p(T_3 - T_4) - c_p(T_2 - T_1)}{q_{\text{in}}} \\ &= \frac{\left(1.0035\ \frac{\text{kJ}}{\text{kg·K}}\right)\left(\begin{array}{r}1220.7\text{K} - 700.1\text{K} \\ - 523.1\text{K} + 300\text{K}\end{array}\right)}{700\ \frac{\text{kJ}}{\text{kg}}} \\ &= 0.4265 \quad (43\%) \end{aligned}$$

The answer is B.

16. $$\text{v}_2 = \frac{Q}{A_2} = \frac{0.03\ \frac{\text{m}^3}{\text{s}}}{\left(\frac{\pi}{4}\right)\left((50\ \text{mm})\left(\frac{1\ \text{m}}{1000\ \text{mm}}\right)\right)^2}$$
$$= 15.28\ \text{m/s}$$

$$\text{v}_1 = \frac{\text{v}_2 A_2}{A_1} = \frac{\text{v}_2 D_2^2}{D_1^2}$$
$$= \frac{\left(15.28\ \frac{\text{m}}{\text{s}}\right)\left((50\ \text{mm})\left(\frac{1\ \text{m}}{1000\ \text{mm}}\right)\right)^2}{(0.1\ \text{m})^2}$$
$$= 3.82\ \text{m/s}$$

$$w = \frac{p_2 - p_1}{\rho} + \frac{\text{v}_2^2 - \text{v}_1^2}{2} + \text{loss}$$
$$= \frac{(600\ \text{kPa})\left(10^3\ \frac{\text{Pa}}{\text{kPa}}\right) - (150\ \text{kPa})\left(10^3\ \frac{\text{Pa}}{\text{kPa}}\right)}{1000\ \frac{\text{kg}}{\text{m}^3}} + \frac{\left(15.28\ \frac{\text{m}}{\text{s}}\right)^2 - \left(3.82\ \frac{\text{m}}{\text{s}}\right)^2}{2} + 200\ \frac{\text{J}}{\text{kg}}$$
$$= 759.4\ \text{J/kg}$$

$$\eta = \frac{w - \text{loss}}{w} = \frac{759.4\ \frac{\text{J}}{\text{kg}} - 200\ \frac{\text{J}}{\text{kg}}}{759.4\ \frac{\text{J}}{\text{kg}}}$$
$$= 0.737 \quad (75\%)$$

The answer is B.

17.

$$\frac{Pw_1}{Pw_2} = \left(\frac{N_1}{N_2}\right)^3 \qquad \textit{Eq. 1}$$
$$\frac{Q_1}{Q_2} = \left(\frac{N_1}{N_2}\right) \qquad \textit{Eq. 2}$$
$$\frac{p_1}{p_2} = \left(\frac{N_1}{N_2}\right)^2 \qquad \textit{Eq. 3}$$

From Eq. 1,

$$\frac{N_1}{N_2} = \left(\frac{Pw_1}{Pw_2}\right)^{\frac{1}{3}} = \left(\frac{1}{3}\right)^{\frac{1}{3}}$$
$$N_2 = N_1\left(\frac{1}{3}\right)^{-\frac{1}{3}} = \left(850\ \frac{\text{rev}}{\text{min}}\right)\left(\frac{1}{3}\right)^{-\frac{1}{3}}$$
$$= 1226\ \text{rpm}$$

From Eq. 2,

$$Q_2 = Q_1\left(\frac{N_2}{N_1}\right) = \left(2.8\ \frac{\text{m}^3}{\text{s}}\right)\left(\frac{1226\ \frac{\text{rev}}{\text{min}}}{850\ \frac{\text{rev}}{\text{min}}}\right)$$
$$= 4.04\ \text{m}^3/\text{s}$$

From Eq. 3,

$$p_2 = p_1\left(\frac{N_2}{N_1}\right)^2 = (5\ \text{cm Hg})\left(\frac{1226\ \frac{\text{rev}}{\text{min}}}{850\ \frac{\text{rev}}{\text{min}}}\right)^2$$
$$= 10.4\ \text{cm Hg}$$

$$\frac{Q_2}{p_2} = \frac{4.04\ \frac{\text{m}^3}{\text{s}}}{10.4\ \text{cm Hg}}$$
$$= 0.388\ \text{m}^3/\text{s}\cdot\text{cm Hg} \quad (0.4\ \text{m}^3/\text{s}\cdot\text{cm Hg})$$

The answer is D.

18. $$\frac{p_2 - p_1}{\rho} + \frac{\text{v}_2^2 - \text{v}_1^2}{2} + g(z_2 - z_1) = w_{\text{comp}} - \text{loss}$$

$$\frac{(800\ \text{kPa})\left(10^3\ \frac{\text{Pa}}{\text{kPa}}\right) - (101\ \text{kPa})\left(10^3\ \frac{\text{Pa}}{\text{kPa}}\right)}{1.23\ \frac{\text{kg}}{\text{m}^3}} + \frac{(0.1)\left(2\ \frac{\text{m}}{\text{s}}\right)^2}{2} + 0 = w_{\text{comp}} - (0.1)(w_{\text{comp}})$$
$$w_{\text{comp}} = 6.31 \times 10^5\ \text{J/kg}$$

$$P_{\text{ideal}} = \dot{m}w_{\text{comp}} = \left(5\ \frac{\text{kg}}{\text{s}}\right)\left(6.31 \times 10^5\ \frac{\text{J}}{\text{kg}}\right)$$
$$= 3.16 \times 10^6\ \text{W} \quad (3.16\ \text{MW})$$

The required power is

$$P_{\text{actual}} = \frac{P_{\text{ideal}}}{\eta} = \frac{3.16\ \text{MW}}{0.85}$$
$$= 3.72\ \text{MW} \quad (3.7\ \text{MW})$$

The answer is C.

19. The force on the plate is equal to its weight.

$$R_y = \dot{m}\text{v} = m_{\text{plate}}g$$
$$\text{v} = \frac{R_y}{\dot{m}} = \frac{m_{\text{plate}}g}{\rho A_0 \text{v}_0}$$
$$= \frac{(1.2\ \text{kg})\left(9.81\ \frac{\text{m}}{\text{s}^2}\right)}{\left(1000\ \frac{\text{kg}}{\text{m}^3}\right)\left(\frac{\pi}{4}\right)(0.02\ \text{m})^2\left(8.5\ \frac{\text{m}}{\text{s}}\right)}$$
$$= 4.41\ \text{m/s}$$

From Bernoulli's equation,

$$\frac{p_0}{\rho g} + \frac{v_0^2}{2g} + z_0 = \frac{p}{\rho g} + \frac{v^2}{2g} + z$$

$$p_0 = p$$

$$z - z_0 = h$$

$$h = \frac{v_0^2 - v^2}{2g} = \frac{\left(8.5\ \frac{\text{m}}{\text{s}}\right)^2 - \left(4.41\ \frac{\text{m}}{\text{s}}\right)^2}{(2)\left(9.81\ \frac{\text{m}}{\text{s}^2}\right)} = 2.69\ \text{m}\quad (2.7\ \text{m})$$

The answer is B.

20. Bernoulli's equation is

$$\frac{p_1}{\rho g} + \frac{v_1^2}{2g} + z_1 = \frac{p_2}{\rho g} + \frac{v_2^2}{2g} + z_2$$

$$z_1 = z_2$$

$$v_2 = v_1\left(\frac{D_1^2}{D_2^2}\right) = \left(0.4\ \frac{\text{m}}{\text{s}}\right)\left(\frac{6\ \text{cm}}{3\ \text{cm}}\right)^2 = 1.6\ \text{m/s}$$

$$\Delta h = \frac{p_1 - p_2}{\rho g} = \frac{v_2^2 - v_1^2}{2g} = \frac{\left(1.6\ \frac{\text{m}}{\text{s}}\right)^2 - \left(0.4\ \frac{\text{m}}{\text{s}}\right)^2}{(2)\left(9.81\ \frac{\text{m}}{\text{s}^2}\right)} = 0.1223\ \text{m}\quad (0.12\ \text{m})$$

The answer is C.

21. Bernoulli's equation is

$$\frac{p_1}{\rho g} + \frac{v_1^2}{2g} + z_1 = \frac{p_2}{\rho g} + \frac{v_2^2}{2g} + z_2$$

However, $p_1 = p_2$, $z_2 - z_1 = h$, and

$$v_2 = \left(\frac{D_1}{D_2}\right)^2 v_1 = \left(\frac{2\ \text{cm}}{8\ \text{cm}}\right)^2\left(2.5\ \frac{\text{m}}{\text{s}}\right) = 0.1563\ \text{m/s}$$

$$h = z_2 - z_1 = \frac{v_1^2 - v_2^2}{2g} = \frac{\left(2.5\ \frac{\text{m}}{\text{s}}\right)^2 - \left(0.1563\ \frac{\text{m}}{\text{s}}\right)^2}{(2)\left(9.81\ \frac{\text{m}}{\text{s}^2}\right)} = 0.3173\ \text{m}\quad (0.32\ \text{m})$$

The answer is B.

22.

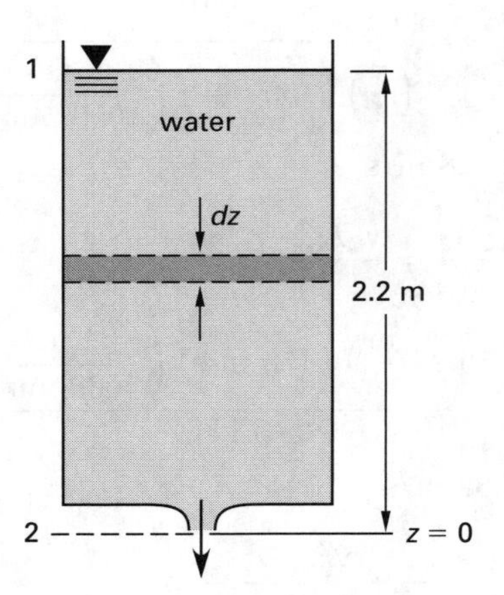

The exit velocity depends on the elevation z in the tank. Applying Bernoulli's equation between points 1 and 2 gives

$$\frac{p_1}{\rho g} + \frac{v_1^2}{2g} + z_1 = \frac{p_2}{\rho g} + \frac{v_2^2}{2g} + z_2$$

However, $p_1 = p_2$, $v_1 \approx 0$, and $z_2 = 0$, so

$$z = \frac{v_2^2}{2g}$$

$$v_2 = \sqrt{2gz}$$

$$Q = A_1\left(-\frac{dz}{dt}\right) = A_2 v_2 = A_2\sqrt{2gz}$$

Separating the variables and integrating gives

$$\int_0^t dt = -\int_{z=2.2}^{0}\left(\frac{A_1}{A_2}\right)\left(\frac{dz}{\sqrt{2gz}}\right) = \left(-\frac{A_1}{A_2\sqrt{2g}}\right)\left(\frac{\sqrt{z}}{\frac{1}{2}}\right)\Bigg|_{2.2\ \text{m}}^{0}$$

$$= \left(\frac{3\ \text{m}^2}{0.02\ \text{m}^2\sqrt{(2)\left(9.81\ \frac{\text{m}}{\text{s}^2}\right)}}\right)\left(\frac{\sqrt{2.2\ \text{m}}}{\frac{1}{2}}\right) = 100.5\ \text{s}\quad (100\ \text{s})$$

The answer is C.

23. To determine the drag coefficient, C_D, the Reynolds number must be determined.

$$\text{Re} = \frac{vD}{\nu} = \frac{\left(22\ \frac{\text{m}}{\text{s}}\right)(10\ \text{cm})\left(\frac{1\ \text{m}}{100\ \text{cm}}\right)}{15.89\times 10^{-6}\ \frac{\text{m}^2}{\text{s}}} = 13.8\times 10^4$$

From a C_D versus Re graph for cylinders in cross flow, $C_D = 1.3$.

The density of air is

$$\rho = \frac{p}{RT} = \frac{(101.3 \text{ kPa})\left(1000 \; \frac{\text{Pa}}{\text{kPa}}\right)}{\left(287 \; \frac{\text{J}}{\text{kg·K}}\right)(300\text{K})}$$

$$= 1.1766 \text{ kg/m}^3$$

The drag force is

$$F_D = C_D \rho \left(\frac{\text{v}^2}{2}\right) A \quad [A \text{ is the projected area.}]$$

$$= (1.3)\left(1.1766 \; \frac{\text{kg}}{\text{m}^3}\right)\left(\frac{\left(22 \; \frac{\text{m}}{\text{s}}\right)^2}{2}\right)$$

$$\times (10 \text{ cm})\left(\frac{1 \text{ m}}{100 \text{ cm}}\right)(1 \text{ m})$$

$$= 37.0 \text{ N} \quad (37 \text{ N})$$

The drag force per unit length is 37 N/m.

The answer is A.

24. The maximum pressure drop corresponds to the maximum velocity in the tube, which in turn corresponds to the maximum Reynolds number for laminar flow.

$$\text{Re}_{\text{max}} = \frac{\text{v}D}{\nu} = 2300$$

$$\nu = \frac{\mu}{\rho} = \frac{855 \times 10^{-6} \; \frac{\text{N·s}}{\text{m}^2}}{1000 \; \frac{\text{kg}}{\text{m}^3}}$$

$$= 0.855 \times 10^{-6} \text{ m}^2/\text{s}$$

$$\text{v} = \text{Re}\left(\frac{\nu}{D}\right)$$

$$= (2300)\left(\frac{0.855 \times 10^{-6} \; \frac{\text{m}^2}{\text{s}}}{(1.5 \text{ mm})\left(\frac{1 \text{ m}}{1000 \text{ mm}}\right)}\right)$$

$$= 1.311 \text{ m/s}$$

For laminar flow, use the Hagen-Poiseuille equation.

$$\Delta p = \frac{128\mu L Q}{\pi D^4} = \frac{32\mu L \text{v}}{D^2}$$

$$= \frac{(32)\left(855 \times 10^{-6} \; \frac{\text{N·s}}{\text{m}^2}\right)(1 \text{ m})\left(1.311 \; \frac{\text{m}}{\text{s}}\right)}{\left((1.5 \text{ mm})\left(\frac{1 \text{ m}}{1000 \text{ mm}}\right)\right)^2}$$

$$= 15.9 \times 10^3 \text{ Pa} \quad (16 \text{ kPa})$$

The pressure drop per unit length is 16 kPa/m.

The answer is C.

25.

$$\frac{\partial T}{\partial x} = -200 + 60x$$

$$q''_{x=0} = -k \left.\frac{\partial T}{\partial x}\right|_{x=0}$$

$$= -\left(2 \; \frac{\text{W}}{\text{m·K}}\right)\left(-200 \; \frac{\text{K}}{\text{m}}\right)$$

$$= 400 \text{ W/m}^2$$

$$q''_{x=0.3 \text{ m}} = -k \left.\frac{\partial T}{\partial x}\right|_{x=0.3 \text{ m}}$$

$$= -\left(2 \; \frac{\text{W}}{\text{m·K}}\right)$$

$$\times \left(-200 \; \frac{\text{K}}{\text{m}} + \left(60 \; \frac{\text{K}}{\text{m}^2}\right)(0.3 \text{ m})\right)$$

$$= 364 \text{ W/m}^2$$

$$q''_{\text{net}} = 400 \; \frac{\text{W}}{\text{m}^2} - 364 \; \frac{\text{W}}{\text{m}^2} = 36 \text{ W/m}^2$$

The answer is C.

26. The outer diameter of the insulation is

$$5 \text{ cm} + 4 \text{ cm} + 4 \text{ cm} = 13 \text{ cm}$$

$$\frac{Q}{L} = \frac{2\pi(T_1 - T_3)}{\frac{\ln \frac{r_2}{r_1}}{k_s} + \frac{\ln \frac{r_3}{r_2}}{k_i}}$$

$$= \frac{2\pi(500\text{K} - 50\text{K})}{\frac{\ln \frac{2.5 \text{ cm}}{1.5 \text{ cm}}}{20 \; \frac{\text{W}}{\text{m·K}}} + \frac{\ln \frac{6.5 \text{ cm}}{2.5 \text{ cm}}}{0.06 \; \frac{\text{W}}{\text{m·K}}}}$$

$$= 177.3 \text{ W/m} \quad (180 \text{ W/m})$$

The answer is D.

27. The Biot number for a sphere is

$$\text{Bi} = \frac{hV}{kA_s} = \frac{h\frac{4}{3}\pi r^3}{k4\pi r^2} = \frac{h\left(\frac{r}{3}\right)}{k}$$

$$= \frac{\left(25 \; \frac{\text{W}}{\text{m}^2\text{·K}}\right)\left(\frac{(4 \text{ cm})\left(\frac{1 \text{ m}}{100 \text{ cm}}\right)}{3}\right)}{380 \; \frac{\text{W}}{\text{m·K}}}$$

$$= 0.00088$$

Since Bi < 0.1, the lumped capacitance method can be used.

$$T - T_\infty = (T_i - T_\infty)e^{-\left(\frac{hA_s}{\rho c_p V}\right)t}$$

$$\frac{T - T_\infty}{T_i - T_\infty} = e^{-\left(\frac{hA_s}{\rho c_p V}\right)t}$$

$$\ln \frac{T - T_\infty}{T_i - T_\infty} = -\frac{hA_s t}{\rho c_p V}$$

$$V = \tfrac{4}{3}\pi r^3 = \frac{\pi D^3}{6}$$

$$t = \left(\frac{\rho\left(\frac{\pi D^3}{6}\right)c_p}{h\pi D^2}\right) \ln \frac{T_i - T_\infty}{T - T_\infty}$$

$$= \left(\frac{\rho D c_p}{6h}\right) \ln \frac{T_i - T_\infty}{T - T_\infty}$$

$$= \left(\frac{\left(8933\ \frac{\text{kg}}{\text{m}^3}\right)(0.08\ \text{m})\left(410\ \frac{\text{J}}{\text{kg}\cdot\text{K}}\right)}{(6)\left(25\ \frac{\text{W}}{\text{m}^2\cdot\text{K}}\right)}\right)$$

$$\times \left(\ln \frac{400°\text{C} - 25°\text{C}}{200°\text{C} - 25°\text{C}}\right)\left(\frac{1\ \text{min}}{60\ \text{s}}\right)$$

$$= 24.8\ \text{min} \quad (25\ \text{min})$$

The answer is D.

28. The maximum Reynolds number is

$$\text{Re} = \frac{\text{v}_\infty L}{\nu}$$

$$= \frac{\left(0.45\ \frac{\text{m}}{\text{s}}\right)(1\ \text{m})}{20.92 \times 10^{-6}\ \frac{\text{m}^2}{\text{s}}}$$

$$= 2.15 \times 10^4$$

Since Re is less than 10^5 the flow over the flat plate is laminar.

$$\text{Nu} = \frac{hL}{k} = 0.648\ \text{Re}_L^{\frac{1}{2}}\text{Pr}^{\frac{1}{3}}$$

$$= (0.648)(2.15 \times 10^4)^{\frac{1}{2}}(0.7)^{\frac{1}{3}}$$

$$= 84.36$$

$$h = \text{Nu}\left(\frac{k}{L}\right)$$

$$= (84.36)\left(\frac{30 \times 10^{-3}\ \frac{\text{W}}{\text{m}\cdot\text{K}}}{1\ \text{m}}\right)$$

$$= 2.53\ \text{W/m}^2\cdot\text{K} \quad (2.5\ \text{W/m}^2\cdot\text{K})$$

The answer is A.

29.

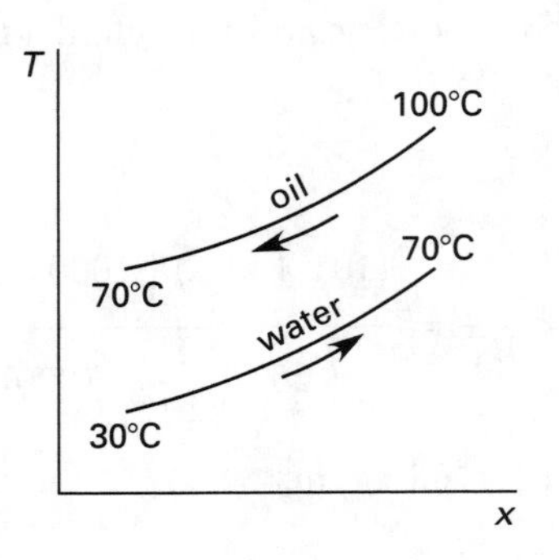

$$Q = \dot{m}_{\text{oil}} c_{p_{\text{oil}}} \Delta T_{\text{oil}}$$

$$= \left(0.1\ \frac{\text{kg}}{\text{s}}\right)\left(1.9\ \frac{\text{kJ}}{\text{kg}\cdot\text{K}}\right)(100°\text{C} - 70°\text{C})$$

$$= 5.7\ \text{kW}$$

$$\Delta T_{\text{lm}} = \frac{(T_{H_o} - T_{C_i}) - (T_{H_i} - T_{C_o})}{\ln \frac{T_{H_o} - T_{C_i}}{T_{H_i} - T_{C_o}}}$$

$$= \frac{(70°\text{C} - 30°\text{C}) - (100°\text{C} - 70°\text{C})}{\ln \frac{70°\text{C} - 30°\text{C}}{100°\text{C} - 70°\text{C}}}$$

$$= 34.76°\text{C} \quad (34.76\text{K})$$

$$Q = UA\ \Delta T_{\text{lm}}$$

$$5.7\ \text{kW} = \left(0.32\ \frac{\text{kW}}{\text{m}^2\cdot\text{K}}\right) A(34.76\text{K})$$

$$A = 0.51\ \text{m}^2 \quad (0.5\ \text{m}^2)$$

The answer is C.

30.

$$\frac{Q_{12}}{A} = \frac{\sigma(T_1^4 - T_2^4)}{\frac{1-\varepsilon_1}{\varepsilon_1} + \frac{1}{F_{12}} + \frac{1-\varepsilon_2}{\varepsilon_2}}$$

$$= \frac{\left(5.67 \times 10^{-8}\ \frac{\text{W}}{\text{m}^2\cdot\text{K}^4}\right) \times \left((700\text{K})^4 - (500\text{K})^4\right)}{\frac{1-0.2}{0.2} + \frac{1}{0.8} + \frac{1-0.4}{0.4}}$$

$$= 1492\ \text{W/m}^2 \quad (1.5\ \text{kW/m}^2)$$

The answer is B.

31. SAE-1080 steel contains 0.8% carbon by weight. From the phase diagram in the NCEES Handbook, the percentage of cementite (Fe_3C) is

$$\text{wt\% Fe}_3\text{C} = \left(\frac{0.80\% - 0.02\%}{6.67\% - 0.02\%}\right) \times 100\%$$

$$= 11.73\%$$

$$\text{vol\% Fe}_3\text{C} = \left(\frac{V_{\text{Fe}_3\text{C}}}{V_{\text{Fe}_3\text{C}} + V_{\text{ferrite}}}\right) \times 100\%$$

$$= \left(\frac{\dfrac{m_{\text{Fe}_3\text{C}}}{\rho_{\text{Fe}_3\text{C}}}}{\dfrac{m_{\text{Fe}_3\text{C}}}{\rho_{\text{Fe}_3\text{C}}} + \dfrac{m_{\text{ferrite}}}{\rho_{\text{ferrite}}}}\right) \times 100\%$$

$$= \left(\frac{(11.73\%)\left(\dfrac{1\text{ g}}{7.66\ \dfrac{\text{g}}{\text{cm}^3}}\right)}{(11.73\%)\left(\dfrac{1\text{ g}}{7.66\ \dfrac{\text{g}}{\text{cm}^3}}\right) + (88.27\%)\left(\dfrac{1\text{ g}}{7.87\ \dfrac{\text{g}}{\text{cm}^3}}\right)}\right) \times 100\%$$

$$= 0.1201 \quad (12\%)$$

The answer is C.

32.

$$\text{Cd} \longrightarrow \text{Cd}^{2+} + 2e^-$$

$$\text{Ni}^{2+} + 2e^- \longrightarrow \text{Ni}$$

$$E_{0,\text{Cd}} = -0.403\text{ V}$$

$$E_{0,\text{Ni}} = -0.250\text{ V}$$

$$E_{\text{Cd}} = -0.403 + 0.0296\log 0.5 = -0.412\text{ V}$$

$$E_{\text{Ni}} = -0.250 + 0.0296\log 0.001 = -0.339\text{ V}$$

Since E_{Cd} has the larger negative potential, Cd will corrode.

$$\Delta E = E_{\text{Cd}} - E_{\text{Ni}} = -0.412\text{ V} - (-0.339\text{ V}) = -0.073\text{ V}$$

The answer is C.

33.

$$\sigma = \frac{F}{A} = \frac{(1.5\text{ kN})\left(10^3\ \dfrac{\text{N}}{\text{kN}}\right)}{\left(\dfrac{\pi}{4}\right)\left((5\text{ mm})\left(\dfrac{1\text{ m}}{1000\text{ mm}}\right)\right)^2} = 7.64 \times 10^7\text{ Pa}$$

This is within the elastic range. The axial strain is

$$\varepsilon_a = \frac{\sigma}{E} = \frac{7.64 \times 10^7\text{ Pa}}{(70\text{ GPa})\left(10^9\ \dfrac{\text{Pa}}{\text{GPa}}\right)} = 0.00109$$

The lateral strain is

$$\varepsilon_l = \nu\varepsilon_a = 0.33\varepsilon_a = \frac{\Delta D}{D_0}$$

$$\Delta D = 0.33\varepsilon_a D_0 = (0.33)(0.00109)(5\text{ mm}) = 0.0018\text{ mm} \quad (0.002\text{ mm})$$

The answer is A.

34.

$$p_A + \gamma_w h_1 - \gamma_{\text{Hg}} h_2 + \gamma_{\text{oil}} h_3 - \gamma_{\text{Hg}} h_4 - \gamma_w h_5 = p_B$$

$$h_1 = 1.0\text{ m} - 0.25\text{ m} = 0.75\text{ m}$$

$$h_2 = 0.7\text{ m} - 0.25\text{ m} = 0.45\text{ m}$$

$$h_3 = 0.7\text{ m} - 0.3\text{ m} = 0.4\text{ m}$$

$$h_4 = 0.85\text{ m} - 0.3\text{ m} = 0.55\text{ m}$$

$$h_5 = 1.75\text{ m} - 0.85\text{ m} = 0.90\text{ m}$$

$$\gamma_w = \rho_w g$$

$$p_A - p_B = \rho_w g(h_5 - h_1) + \gamma_{\text{Hg}}(h_2 + h_4) - \gamma_{\text{oil}} h_3$$

$$= \left(1000\ \frac{\text{kg}}{\text{m}^3}\right)\left(9.81\ \frac{\text{m}}{\text{s}^2}\right) \times \left(\begin{array}{l}(0.90\text{ m} - 0.75\text{ m}) \\ \quad + (13.59)(0.45\text{ m} + 0.55\text{ m}) \\ \quad - (0.873)(0.4\text{ m})\end{array}\right)$$

$$= 1.314 \times 10^5\text{ Pa} \quad (130\text{ kPa})$$

The answer is C.

35. Use the pitot tube equation.

$$\text{v}_2 = \sqrt{2g\left(\frac{\rho_{\text{Hg}}}{\rho_w} - 1\right)h}$$

$$= \sqrt{(2)\left(9.81\ \frac{\text{m}}{\text{s}^2}\right)\left(\frac{13.59}{1} - 1\right) \times (6\text{ cm})\left(\frac{1\text{ m}}{100\text{ cm}}\right)}$$

$$= 3.85\text{ m/s}$$

$$Q = A\text{v}_2 = \left(\frac{\pi}{4}\right)\left((10\text{ cm})\left(\frac{1\text{ m}}{100\text{ cm}}\right)\right)^2\left(3.85\ \frac{\text{m}}{\text{s}}\right)$$

$$= 3.02 \times 10^{-2}\text{ m}^3\text{/s} \quad (0.03\text{ m}^3\text{/s})$$

The answer is A.

36.

$$\Delta p = (\rho_{Hg} - \rho_w)gh = \rho_w(\mathrm{SG}_{Hg} - 1)gh$$

$$\begin{aligned}\frac{\Delta p}{\rho_w g} &= (\mathrm{SG}_{Hg} - 1)h \\ &= (13.59 - 1)\left((20\ \mathrm{cm})\left(\frac{1\ \mathrm{m}}{100\ \mathrm{cm}}\right)\right) \\ &= 2.518\ \mathrm{m}\end{aligned}$$

$$\begin{aligned}Q &= C_d A_t \sqrt{\frac{2g\left(\frac{\Delta p}{\rho_w g}\right)}{1 - \left(\frac{A_t}{A}\right)^2}} \\ &= C_d\left(\frac{\pi}{4}\right)D_t^2\sqrt{\frac{2g\left(\frac{\Delta p}{\rho_w g}\right)}{1 - \left(\frac{D_t}{D}\right)^4}}\end{aligned}$$

$$0.0045\ \frac{\mathrm{m}^3}{\mathrm{s}} = C_d\left(\frac{\pi}{4}\right)\left((0.03\ \mathrm{cm})\left(\frac{1\ \mathrm{m}}{100\ \mathrm{cm}}\right)\right)^2 \times \sqrt{\frac{(2)\left(9.81\ \frac{\mathrm{m}}{\mathrm{s}^2}\right)(2.518\ \mathrm{m})}{1 - \left(\frac{(3\ \mathrm{cm})\left(\frac{1\ \mathrm{m}}{100\ \mathrm{cm}}\right)}{(10\ \mathrm{cm})\left(\frac{1\ \mathrm{m}}{100\ \mathrm{cm}}\right)}\right)^4}}$$

$$C_d = 0.902 \quad (0.90)$$

The answer is A.

37.

$$\mathrm{v} = \frac{Q}{\left(\frac{\pi}{4}\right)D^2} = \frac{0.03\ \frac{\mathrm{m}^3}{\mathrm{s}}}{\left(\frac{\pi}{4}\right)\left((8\ \mathrm{cm})\left(\frac{1\ \mathrm{m}}{100\ \mathrm{cm}}\right)\right)^2} = 5.968\ \mathrm{m/s}$$

$$\mathrm{Re} = \frac{\mathrm{v}D}{\nu} = \frac{\left(5.968\ \frac{\mathrm{m}}{\mathrm{s}}\right)(0.08\ \mathrm{m})}{9.609 \times 10^{-5}\ \frac{\mathrm{m}^2}{\mathrm{s}}} = 4969 \quad [\text{turbulent}]$$

From the NCEES Handbook, $C = 0.61$.

$$Q = CA\sqrt{\frac{2\Delta p}{\rho}}$$

$$0.03\ \frac{\mathrm{m}^3}{\mathrm{s}} = (0.61)\left(\frac{\pi}{4}\right)\left((5\ \mathrm{cm})\left(\frac{1\ \mathrm{m}}{100\ \mathrm{cm}}\right)\right)^2 \times \sqrt{\frac{2\Delta p}{1000\ \frac{\mathrm{kg}}{\mathrm{m}^3}}}$$

$$\Delta p = 313.6 \times 10^3\ \mathrm{Pa} \quad (310\ \mathrm{kPa})$$

The answer is C.

38. For dynamic similarity, the two Reynolds numbers must be the same.

$$\mathrm{Re} = \frac{D\mathrm{v}}{\nu}$$

$$\mathrm{Re}_{oil} = \mathrm{Re}_{water}$$

$$\frac{(30\ \mathrm{cm})\left(\frac{1\ \mathrm{m}}{100\ \mathrm{cm}}\right) \times \left(12.5\ \frac{\mathrm{m}}{\mathrm{s}}\right)}{1120 \times 10^{-6}\ \frac{\mathrm{m}^2}{\mathrm{s}}} = \frac{(2.5\ \mathrm{cm})\left(\frac{1\ \mathrm{m}}{100\ \mathrm{cm}}\right) \times \mathrm{v}_{water}}{1.08 \times 10^{-6}\ \frac{\mathrm{m}^2}{\mathrm{s}}}$$

$$\mathrm{v}_{water} = 0.145\ \mathrm{m/s} \quad (0.15\ \mathrm{m/s})$$

The answer is A.

39.

$$\mathrm{M} = \frac{\mathrm{v}}{a}$$

$$\begin{aligned}\mathrm{v} &= \mathrm{M}a = \mathrm{M}\sqrt{kRT} \\ &= 1.5\sqrt{(1.4)\left(287\ \frac{\mathrm{J}}{\mathrm{kg{\cdot}K}}\right)(320\mathrm{K})} \\ &= 537.9\ \mathrm{m/s} \quad (540\ \mathrm{m/s})\end{aligned}$$

The answer is D.

40.

$$\begin{aligned}A_{tube} &= (4\ \mathrm{cm}^2)\left(\frac{1\ \mathrm{m}}{100\ \mathrm{cm}}\right)^2 \\ &= 4 \times 10^{-4}\ \mathrm{m}^2\end{aligned}$$

$$\begin{aligned}A_{bolt} &= (2\ \mathrm{cm}^2)\left(\frac{1\ \mathrm{m}}{100\ \mathrm{cm}}\right)^2 \\ &= 2 \times 10^{-4}\ \mathrm{m}^2\end{aligned}$$

The thread pitch is $h = 2$ mm (0.002 m). The displacement of the bolt relative to the tube, is

$$\delta_{rel} = \delta_{bolt} - \delta_{tube} = \frac{h}{4} \qquad \textit{Eq. 1}$$

$$\delta_{tube} = \frac{F_{tube}L_{tube}}{A_{tube}E_{tube}} \qquad \textit{Eq. 2}$$

$$\delta_{bolt} = \frac{F_{bolt}L_{bolt}}{A_{bolt}E_{bolt}} \qquad \textit{Eq. 3}$$

Combine Eqs. 1, 2, and 3.

$$\frac{h}{4} = \frac{F_{bolt}L_{bolt}}{A_{bolt}E_{bolt}} - \frac{F_{tube}L_{tube}}{A_{tube}E_{tube}}$$

Since there are no external forces,

$$-F_{\text{tube}} = F_{\text{bolt}} \quad \text{[tube in compression, bolt in tension]}$$

The moduli of elasticity for steel, E_{bolt}, and aluminum, E_{tube}, can be found from tabulated data.

Solve for F_{bolt}.

$$F_{\text{bolt}} = \frac{\frac{h}{4}}{\frac{L_{\text{bolt}}}{A_{\text{bolt}}E_{\text{bolt}}} + \frac{L_{\text{tube}}}{A_{\text{tube}}E_{\text{tube}}}}$$

$$= \frac{\frac{0.002 \text{ m}}{4}}{\frac{0.7 \text{ m}}{(2 \times 10^{-4} \text{ m}^2)(205 \times 10^9 \text{ Pa})} + \frac{0.7 \text{ m}}{(4 \times 10^{-4} \text{ m}^2)(70 \times 10^9 \text{ Pa})}}$$

$$= 11\,884 \text{ N}$$

The stress in the bolt is

$$\sigma = \frac{F}{A} = \frac{11\,884 \text{ N}}{(2 \text{ cm}^2)\left(\frac{1 \text{ m}}{100 \text{ cm}}\right)^2}$$

$$= 5.94 \times 10^7 \text{ Pa} \quad (60 \text{ MPa})$$

The answer is C.

41.

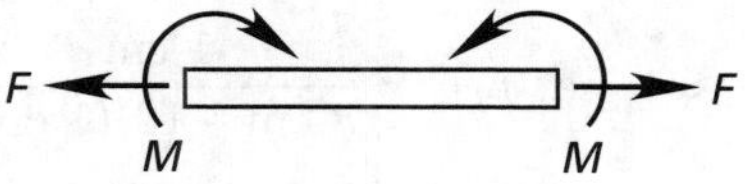

$$x = 10 \text{ cm} - \frac{6 \text{ cm}}{2}$$

$$= 7 \text{ cm} \quad (0.07 \text{ m})$$

$$M = Fx = F(0.07 \text{ m})$$

$$I = \frac{\pi}{4}R^4 = \frac{\pi}{4}\left(\frac{(6 \text{ cm})\left(\frac{1 \text{ m}}{100 \text{ cm}}\right)}{2}\right)^4$$

$$= 6.36 \times 10^{-7} \text{ m}^4$$

$$\sigma = \frac{F}{A} + \frac{Mc}{I}$$

$$(1.2 \text{ MPa})\left(10^6 \frac{\text{Pa}}{\text{MPa}}\right) = \frac{F}{\left(\frac{\pi}{4}\right)\left((6 \text{ cm})\left(\frac{1 \text{ m}}{100 \text{ cm}}\right)\right)^2} + \frac{(0.07F \text{ m})\left(\frac{(6 \text{ cm}) \times \left(\frac{1 \text{ m}}{100 \text{ cm}}\right)}{2}\right)}{6.36 \times 10^{-7} \text{ m}^4}$$

$$F = 328.3 \text{ N} \quad (330 \text{ N})$$

The answer is C.

42. The elongation due to temperature is

$$\delta_t = \alpha \Delta T L$$

$$= \left(15 \times 10^{-6} \frac{1}{^\circ\text{C}}\right)(500^\circ\text{C}) \times (75 \text{ cm})\left(\frac{1 \text{ m}}{100 \text{ cm}}\right)$$

$$= 0.00563 \text{ m} \quad (5.63 \text{ mm})$$

$$P_{\text{cr}} = \frac{\pi^2 EI}{k^2L^2}$$

$$= \frac{\pi^2\left((150 \text{ GPa})\left(10^9 \frac{\text{Pa}}{\text{GPa}}\right)\right) \times \left(\frac{\pi}{64}\right)\left((1 \text{ cm})\left(\frac{1 \text{ m}}{100 \text{ cm}}\right)\right)^4}{(1.0)^2\left((0.75)\left(\frac{1 \text{ m}}{100 \text{ cm}}\right)\right)^2}$$

$$= 1292 \text{ N}$$

$$\delta_{\text{comp}} = \frac{PL}{EA}$$

$$= \frac{(1292 \text{ N})\left((0.75)\left(\frac{1 \text{ m}}{100 \text{ cm}}\right)\right)}{(150 \text{ GPa})\left(10^9 \frac{\text{Pa}}{\text{GPa}}\right)\left(\frac{\pi}{4}\right) \times \left((1 \text{ cm})\left(\frac{1 \text{ m}}{1000 \text{ cm}}\right)\right)^2}$$

$$= 8.23 \times 10^{-5} \text{ m} \quad (0.0823 \text{ mm})$$

$$\delta = \delta_T - \delta_{\text{comp}}$$

$$= 5.63 \text{ mm} - 0.0823 \text{ mm}$$

$$= 5.55 \text{ mm}$$

The answer is C.

43. The moment at the centroid of the bolt group is

$$\begin{aligned} M &= Fd \\ &= (20\ \text{kN})\left((250\ \text{mm})\left(\frac{1\ \text{m}}{1000\ \text{mm}}\right) + \frac{(150\ \text{mm})\left(\frac{1\ \text{m}}{1000\ \text{mm}}\right)}{2}\right) \\ &= 6.5\ \text{kN·m} \end{aligned}$$

The distance from the centroid to the center of each bolt is

$$\begin{aligned} r &= \sqrt{(75\ \text{mm})^2 + (75\ \text{mm})^2} \\ &= 106\ \text{mm} \quad (0.106\ \text{m}) \end{aligned}$$

The shear load per critical bolt is the sum of the direct and torsional shears.

The direct shear per bolt is

$$\begin{aligned} V_{F,y} &= \frac{F}{4} = \frac{20\ \text{kN}}{4} \\ &= 5\ \text{kN} \quad [\text{vertical}] \end{aligned}$$

The torsional shear per bolt is

$$\begin{aligned} V_M &= \frac{M}{4r} = \frac{6.5\ \text{kN·m}}{(4)(0.106\ \text{m})} \\ &= 15.33\ \text{kN} \end{aligned}$$

The torsional shear is perpendicular to a line directed from the centroid of the bolt group to the bolt. For bolt D, the vertical and horizontal components are

$$\begin{aligned} V_{M,y} &= V_M \sin 45^\circ \\ &= (15.33\ \text{kN}) \sin 45^\circ \\ &= 10.84\ \text{kN} \\ V_{M,x} &= 10.84\ \text{kN} \end{aligned}$$

The resultant shear on the bolt is

$$\begin{aligned} V &= \sqrt{(V_{F,y} + V_{M,y})^2 + V_{M,x}^2} \\ &= \sqrt{(5\ \text{kN} + 10.84\ \text{kN})^2 + (10.84\ \text{kN})^2} \\ &= 19.2\ \text{kN} \quad (19\ \text{kN}) \end{aligned}$$

The answer is C.

44.

$$\begin{aligned} J &= \left(\frac{\pi}{32}\right)(D_o^4 - D_i^4) \\ &= \left(\frac{\pi}{32}\right)\left((40\ \text{mm})^4 - (30\ \text{mm})^4\right) \\ &= 1.718 \times 10^5\ \text{mm}^4 \quad (1.718 \times 10^{-7}\ \text{m}^4) \end{aligned}$$

$$\begin{aligned} P &= T\omega = 2\pi T\left(\frac{N}{60\ \frac{\text{s}}{\text{min}}}\right) \\ T &= \frac{\left(60\ \frac{\text{s}}{\text{min}}\right)P}{2\pi N} \\ &= \frac{\left(60\ \frac{\text{s}}{\text{min}}\right)\left((35\ \text{kW})\left(1000\ \frac{\text{W}}{\text{kW}}\right)\right)}{2\pi\left(1000\ \frac{\text{rev}}{\text{min}}\right)} \\ &= 334.2\ \text{N·m} \\ \tau &= \frac{Tc}{J} \\ &= \frac{(334.2\ \text{N·m})\left(\frac{(40)\left(\frac{1\ \text{m}}{1000\ \text{mm}}\right)}{2}\right)}{1.718 \times 10^{-7}\ \text{m}^4} \\ &= 3.89 \times 10^7\ \text{Pa} \quad (40\ \text{MPa}) \end{aligned}$$

The answer is A.

45. The allowable force is

$$P = \frac{\sigma_{\max} A_{\text{net}}}{(\text{FS})K_t}$$

From the stress concentration factor diagram,

$$\text{hole: } K_t \approx 2.0 \text{ at } r/d = \frac{(3\ \text{cm})(2)}{(2)(8\ \text{cm} - 3\ \text{cm})} = 0.6$$

$$\text{fillet: } K_t \approx 1.7 \text{ at } r/d = \frac{1\ \text{cm}}{8\ \text{cm} - (2)(1\ \text{cm})} = 0.17$$

$$\begin{aligned} \text{hole: } P &= \frac{(1.5\ \text{MPa})\left(10^6\ \frac{\text{Pa}}{\text{MPa}}\right) \times (0.01\ \text{m})(0.05\ \text{m})}{(2.5)(2.0)} \\ &= 150\ \text{N} \end{aligned}$$

$$\begin{aligned} \text{fillet: } P &= \frac{(1.5\ \text{MPa})\left(10^6\ \frac{\text{Pa}}{\text{MPa}}\right) \times (0.01\ \text{m})(0.06\ \text{m})}{(2.5)(1.7)} \\ &= 212\ \text{N} \\ &= 150\ \text{N} \quad [\text{limited by the hole}] \end{aligned}$$

The answer is C.

46.

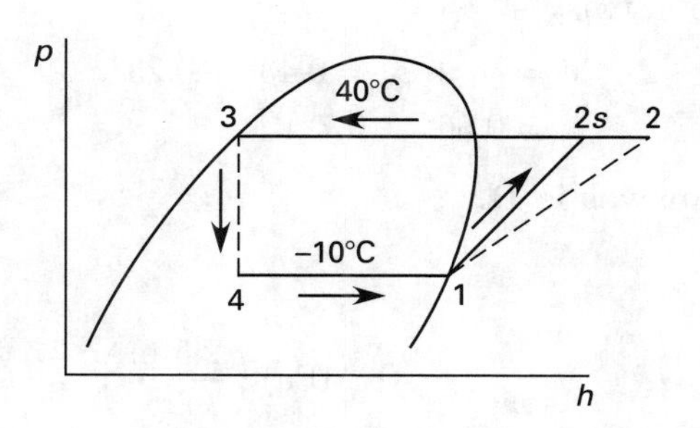

The efficiency of the compressor is

$$\eta_{\text{comp}} = \frac{h_{2s} - h_1}{h_2 - h_1}$$

From R-134a tables or the *p*-*h* diagram, at −10°C,

$$h_1 = 392.9 \text{ kJ/kg}$$
$$s_1 = 1.7341 \text{ kJ/kg·K}$$

At 40°C,

$$p_{\text{sat}} = 1.017 \text{ MPa}$$

$s_{2s} = s_1 = 1.7341$ kJ/kg·K and $p_{2s} = 1.017$ MPa,

$$\begin{aligned} h_{2s} &\approx 426.4 \text{ kJ/kg·K} \\ w = h_2 - h_1 &= \frac{h_{2s} - h_1}{\eta_{\text{comp}}} \\ &= \frac{426.4 \ \frac{\text{kJ}}{\text{kg}} - 392.9 \ \frac{\text{kJ}}{\text{kg}}}{0.8} \\ &= 41.88 \text{ kJ/kg} \quad (42 \text{ kJ/kg}) \end{aligned}$$

The answer is B.

47. The inlet and exit enthalpies are found from R-134a tables.

At 0.8 MPa and 70°C: $h_i = 455.3$ kJ/kg
At 0.8 MPa: $h_e = 243.7$ kJ/kg [sat]

The inlet and exit enthalpies of water are found from steam tables.

At 20°C: $h_i = 83.95$ kJ/kg [sat]
At 30°C: $h_e = 125.78$ kJ/kg [sat]

The energy balance is

$$\dot{m}_{\text{R-134}}(h_i - h_e)_{\text{R-134}} = \dot{m}_{\text{water}}(h_e - h_i)_{\text{water}}$$

$$\begin{aligned} &\left(0.1 \ \frac{\text{kg}}{\text{s}}\right)\left(455.3 \ \frac{\text{kJ}}{\text{kg}} - 243.7 \ \frac{\text{kJ}}{\text{kg}}\right) \\ &\quad = \dot{m}_{\text{water}}\left(125.78 \ \frac{\text{kJ}}{\text{kg}} - 83.95 \ \frac{\text{kJ}}{\text{kg}}\right) \\ &\dot{m}_{\text{water}} = 0.506 \text{ kg/s} \quad (0.51 \text{ kg/s}) \end{aligned}$$

The answer is A.

48. Let x_6 represent the quality at state 6.

$$\begin{aligned} h_6 &= h_7 + x_6(h_3 - h_7) \\ x_6 &= \frac{h_6 - h_7}{h_3 - h_7} = \frac{243.7 \ \frac{\text{kJ}}{\text{kg}} - 200.9 \ \frac{\text{kJ}}{\text{kg}}}{399.2 \ \frac{\text{kJ}}{\text{kg}} - 200.9 \ \frac{\text{kJ}}{\text{kg}}} \\ &= 0.2158 \\ q_L &= (1 - x_6)(h_1 - h_8) \\ &= (1 - 0.2158)\left(388.5 \ \frac{\text{kJ}}{\text{kg}} - 200.9 \ \frac{\text{kJ}}{\text{kg}}\right) \\ &= 147.1 \text{ kJ/kg} \quad (145 \text{ kJ/kg}) \quad \text{[entering condenser]} \end{aligned}$$

The answer is C.

49. The horizontal component of the force is

$$\begin{aligned} F_h &= F \cos 30° = (30 \text{ kN}) \cos 30° \\ &= 25.98 \text{ kN} \end{aligned}$$

The shear area required is

$$\begin{aligned} A_s &= \frac{F_h}{S_{\text{allowable}}} = \frac{25.98 \text{ kN}}{600 \text{ kPa}} \\ &= 0.0433 \text{ m}^2 \quad (43\,300 \text{ mm}^2) \end{aligned}$$

The required width is

$$\begin{aligned} w &= \frac{A_s}{200 \text{ mm}} = \frac{43\,300 \text{ mm}^2}{200 \text{ mm}} \\ &= 217 \text{ mm} \quad (220 \text{ mm}) \end{aligned}$$

The answer is D.

50. The force along each bar is

$$\begin{aligned} F_b &= \frac{F}{(2) \cos 60°} = \frac{90 \text{ kN}}{(2)(0.5)} \\ &= 90 \text{ kN} \end{aligned}$$

The strain energy of the two bars is

$$U = (2)\left(\frac{F_b^2 L}{2EA}\right)$$

The work done by the vertical load is equal to the strain energy.

$$W = \frac{F_b \delta_B}{2} = U$$

$$\delta_B = \frac{2F_b L}{EA}$$

$$= \frac{(2)(90 \text{ kN})\left(1000 \ \frac{\text{N}}{\text{kN}}\right)(1 \text{ m})}{(210 \times 10^9 \text{ Pa})(4 \text{ cm}^2)\left(\frac{1 \text{ m}}{100 \text{ cm}}\right)^2 (0.5)^2}$$

$$= 8.57 \times 10^{-3} \text{ m} \quad (8.6 \text{ mm})$$

The answer is D.

51. The thermal elongation equals the mechanical compression.

$$0 = \alpha \Delta T L + \frac{FL}{AE}$$

$$\sigma = \frac{F}{A} = -\alpha \Delta T E$$

$$= -\left(23 \times 10^{-6} \ \frac{1}{°\text{C}}\right)(100°\text{C} - 25°\text{C})(70 \times 10^9 \text{ Pa})$$

$$= -1.21 \times 10^8 \text{ Pa} \quad (120 \text{ MPa}) \quad \text{[compressive]}$$

The answer is B.

52.

$$T = T_1 + T_2$$
$$= 3000 \text{ N·m} + 1000 \text{ N·m}$$
$$= 4000 \text{ N·m}$$

$$\phi_1 = \frac{TL_1}{GJ_1} = \frac{TL_1}{G\left(\frac{\pi}{32}\right)D^4}$$

$$= \frac{(4000 \text{ N·m})(0.6 \text{ m})}{(80 \text{ GPa})\left(10^9 \ \frac{\text{Pa}}{\text{GPa}}\right)\left(\frac{\pi}{32}\right) \times \left((80 \text{ mm})\left(\frac{1 \text{ m}}{1000 \text{ mm}}\right)\right)^4}$$

$$= 0.0075 \text{ rad} \quad (0.43°)$$

$$\phi_2 = \frac{T_2 L_2}{GJ_2}$$

$$= \frac{(1000 \text{ N·m})(0.4 \text{ m})}{(80 \times 10^9 \text{ Pa})\left(\frac{\pi}{32}\right) \times \left((60 \text{ mm})\left(\frac{1 \text{ m}}{1000 \text{ mm}}\right)\right)^4}$$

$$= 0.0039 \text{ rad} \quad (0.23°)$$

The total twist is

$$\phi = \phi_1 + \phi_2 = 0.43° + 0.23°$$
$$= 0.66° \quad (0.7°)$$

The answer is D.

53.

$$\sigma_1 = \frac{pr}{t} = \frac{(8 \text{ MPa})\left(\frac{2.5 \text{ m}}{2}\right)}{(15 \text{ mm})\left(\frac{1 \text{ m}}{1000 \text{ mm}}\right)}$$
$$= 666.67 \text{ MPa}$$

$$\sigma_2 = \frac{pr}{2t} = \frac{\sigma_1}{2}$$
$$= \frac{666.67 \text{ MPa}}{2}$$
$$= 333.33 \text{ MPa}$$

$$\sigma_x = \sigma_2 = 333.33 \text{ MPa}$$

The normal stress along the seam is

$$\sigma_{x'} = \frac{\sigma_x + \sigma_y}{2} + \left(\frac{\sigma_x - \sigma_y}{2}\right)\cos 2\theta + \tau_{xy} \sin 2\theta$$

$$\tau_{xy} = 0$$

$$\theta = 90° + 45° = 135°$$

$$\sigma_{x'} = \frac{666.67 \text{ MPa} + 333.33 \text{ MPa}}{2} + \left(\frac{\sigma_x - \sigma_y}{2}\right)(0) + 0$$

$$= 500 \text{ MPa}$$

The answer is A.

54.

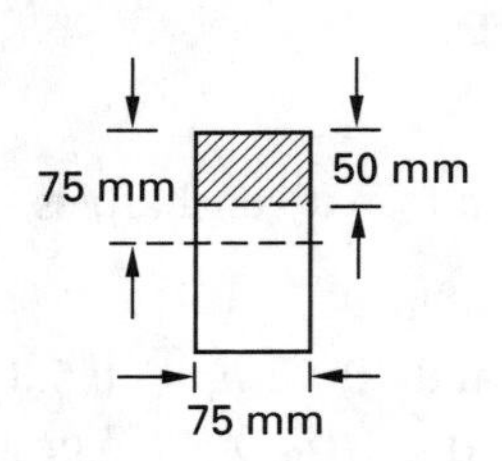

$$I = \frac{bh^3}{12} = \frac{(75 \text{ mm})\left(\frac{1 \text{ m}}{1000 \text{ mm}}\right) \times \left((150 \text{ mm})\left(\frac{1 \text{ m}}{1000 \text{ mm}}\right)\right)^3}{12}$$
$$= 2.11 \times 10^{-5} \text{ m}^4$$

$$M_P = Fd = (150 \text{ kN})(75 \text{ mm})\left(\frac{1 \text{ m}}{1000 \text{ mm}}\right)$$
$$= 112.5 \text{ kN·m}$$

The distance from the neutral axis to P is

$$c = 75 \text{ mm} - 50 \text{ mm} = 25 \text{ mm} \quad (0.025 \text{ m})$$

$$\sigma_x = \frac{Mc}{I} = \frac{(112.5 \text{ kN·m})\left(1000 \ \frac{\text{N}}{\text{kN}}\right)(0.025 \text{ m})}{2.11 \times 10^{-5} \text{ m}^4} = 1.333 \times 10^8 \text{ Pa} \quad (133.3 \text{ MPa})$$

The shear force is

$$V = 150 \text{ kN}$$

The first moment of the shaded area about the neutral axis is Q.

$$\tau = \frac{VQ}{Ib} = \tau_{xy}$$

$$= \frac{(150 \text{ kN})\left(1000 \ \frac{\text{N}}{\text{kN}}\right)(75 \text{ mm})\left(\frac{1 \text{ m}}{1000 \text{ mm}}\right) \times (50 \text{ mm})\left(\frac{1 \text{ m}}{1000 \text{ mm}}\right) \times (50 \text{ mm})\left(\frac{1 \text{ m}}{1000 \text{ mm}}\right)}{(2.11 \times 10^{-5} \text{ m}^4)(75 \text{ mm})\left(\frac{1 \text{ m}}{1000 \text{ mm}}\right)}$$

$$= 1.777 \times 10^7 \text{ Pa} \quad (17.77 \text{ MPa})$$

$$\tan 2\theta = \frac{2\tau_{xy}}{\sigma_x - \sigma_y} = \frac{(2)(17.77 \text{ MPa})}{133.3 \text{ MPa} - 0} = 0.267$$

$$2\theta = \tan^{-1} 0.267 = 14.93° \text{ or } 194.93°$$

$$\sigma_{x_1} = \frac{\sigma_x + \sigma_y}{2} + \left(\frac{\sigma_x - \sigma_y}{2}\right)\cos 2\theta + \tau_{xy} \sin 2\theta$$

For $2\theta = 14.93°$,

$$\sigma_{x_1} = \frac{133.3 \text{ MPa} + 0}{2} + \left(\frac{133.3 \text{ MPa} - 0}{2}\right) \times \cos 14.93° + (17.77 \text{ MPa}) \sin 14.93° = 135.6 \text{ MPa}$$

For $2\theta = 194.93°$,

$$\frac{133.3 \text{ MPa} + 0}{2} + \left(\frac{133.3 \text{ MPa} - 0}{2}\right)\cos 194.93° + (17.77 \text{ MPa}) \sin 194.93° = -2.33 \text{ MPa}$$

$$\left|\frac{\sigma_1}{\sigma_2}\right| = \left|\frac{\tau_{x1}}{\tau_{x2}}\right| = \left|\frac{135.6 \text{ MPa}}{-2.33 \text{ MPa}}\right| = 58.2 \quad (60)$$

The answer is B.

55. The initial enthalpy at 100°C is

$$h_1 = h_f + xh_{fg} = 419.04 \ \frac{\text{kJ}}{\text{kg}} + (0.3)\left(2257.0 \ \frac{\text{kJ}}{\text{kg}}\right) = 1096.1 \text{ kJ/kg}$$

The volume of the rigid container is

$$V = V_{\text{liquid}} + V_{\text{vapor}} = m((1-x)v_f + xv_g)$$

$$= (1.0 \text{ kg})\left((1 - 0.3)\left(0.001044 \ \frac{\text{m}^3}{\text{kg}}\right) + (0.3)\left(1.6729 \ \frac{\text{m}^3}{\text{kg}}\right)\right) = 0.5026 \text{ m}^3$$

When all of the liquid has been vaporized, the specific volume will be

$$v_{g,2} = \frac{V}{m} = \frac{0.5026 \text{ m}^3}{1.0 \text{ kg}} = 0.5026 \text{ m}^3/\text{kg}$$

Use the steam tables for this value of v_g.

$$T_2 \approx 140°\text{C}$$

$$h_{g,2} = 2733.9 \text{ kJ/kg}$$

The energy required is

$$m(h_{g,2} - h_1) = (1.0 \text{ kg})\left(2733.9 \ \frac{\text{kJ}}{\text{kg}} - 1096.1 \ \frac{\text{kJ}}{\text{kg}}\right) = 1637.8 \text{ kJ} \quad (1.6 \text{ MJ})$$

Notice that energy is required to raise the pressure as well as the temperature, in addition to changing the phase. Enthalpy ($h = u + pv$) accounts for all of these factors.

The answer is D.

56.

$$m_1 = \frac{p_1 V_1}{RT_1} = \frac{(150 \text{ kPa})\left(1000 \ \frac{\text{Pa}}{\text{kPa}}\right)(1 \text{ m}^3)}{\left(287 \ \frac{\text{J}}{\text{kg·K}}\right)(300\text{K})} = 1.74 \text{ kg}$$

$$m_2 = \frac{p_2 V_2}{RT_2} = \frac{(750 \text{ kPa})\left(1000 \ \frac{\text{Pa}}{\text{kPa}}\right)(1 \text{ m}^3)}{\left(287 \ \frac{\text{J}}{\text{kg·K}}\right)(300\text{K})} = 8.71 \text{ kg}$$

$$m_{\text{entering}} = m_2 - m_1 = 8.71 \text{ kg} - 1.74 \text{ kg} = 6.97 \text{ kg}$$

The first law of thermodynamics for this unsteady flow process is

$$Q + m_{\text{in}} h_{\text{in}} = m_2 u_2 - m_1 u_1$$

Substituting in the first law and assuming air to be an ideal gas,

$$\begin{aligned} Q &= -m_{\text{in}} c_p T_{\text{in}} + m_2 c_v T_2 - m_1 c_v T_1 \\ &= -(6.97 \text{ kg})\left(1.0035 \ \frac{\text{kJ}}{\text{kg·K}}\right)(300\text{K}) \\ &\quad + (8.71 \text{ kg})\left(0.7165 \ \frac{\text{kJ}}{\text{kg·K}}\right)(300\text{K}) \\ &\quad - (1.74 \text{ kg})\left(0.7165 \ \frac{\text{kJ}}{\text{kg·K}}\right)(300\text{K}) \\ &= -600 \text{ kJ} \quad \text{[to the surroundings]} \end{aligned}$$

The answer is D.

57. The energy balance equation gives

$$m_{\text{Cu}} c_{\text{Cu}} \Delta T_{\text{Cu}} = -m_{\text{water}} c_{\text{water}} \Delta T_{\text{water}}$$

$$(5 \text{ kg})\left(0.39 \ \frac{\text{kJ}}{\text{kg·K}}\right)(T_f - 200°\text{C}) = -(10 \text{ kg})\left(4.18 \ \frac{\text{kJ}}{\text{kg·K}}\right)(T_f - 25°\text{C})$$

$$T_f = 32.8°\text{C}$$

$$\begin{aligned} \Delta S_{\text{Cu}} &= m_{\text{Cu}} c_{\text{Cu}} \ln\left(\frac{T_f}{T_i}\right)_{\text{Cu}} \\ &= (5 \text{ kg})\left(0.39 \ \frac{\text{kJ}}{\text{kg·K}}\right)\ln\frac{32.8°\text{C} + 273°}{200°\text{C} + 273°} \\ &= -0.8505 \text{ kJ/K} \end{aligned}$$

$$\begin{aligned} \Delta S_{\text{water}} &= m_{\text{water}} c_{\text{water}} \ln\left(\frac{T_f}{T_i}\right)_{\text{water}} \\ &= (10 \text{ kg})\left(4.18 \ \frac{\text{kJ}}{\text{kg·K}}\right)\ln\frac{32.8°\text{C} + 273°}{25°\text{C} + 273°} \\ &= 1.0800 \text{ kJ/K} \end{aligned}$$

$$\begin{aligned} S_{\text{generated}} &= -0.8505 \ \frac{\text{kJ}}{\text{K}} + 1.0800 \ \frac{\text{kJ}}{\text{K}} \\ &= 0.2295 \text{ kJ/K} \quad (0.23 \text{ kJ/K}) \end{aligned}$$

The answer is A.

58. For an ideal gas in an adiabatic process with no change in kinetic and potential energies,

$$w_{\text{in}} = h_e - h_i = c_p(T_e - T_i)$$

$$150 \ \frac{\text{kJ}}{\text{kg}} = \left(1.0035 \ \frac{\text{kJ}}{\text{kg·°C}}\right)(T_e - 25°\text{C})$$

$$T_e = 174.5°\text{C}$$

The change of entropy per unit mass for the process is

$$\begin{aligned} \Delta s &= c_p \ln\frac{T_e}{T_i} - R \ln\frac{p_e}{p_i} \\ &= \left(1.0035 \ \frac{\text{kJ}}{\text{kg·K}}\right)\ln\frac{174.5°\text{C} + 273°}{25°\text{C} + 273°} \\ &\quad - \left(0.287 \ \frac{\text{kJ}}{\text{kg·K}}\right)\ln\frac{300 \text{ kPa}}{100 \text{ kPa}} \\ &= 0.0927 \text{ kJ/kg·K} \quad (0.09 \text{ kJ/kg·K}) \end{aligned}$$

The answer is C.

59.

$$h_i = h_e + \frac{\text{v}_e^2}{2}$$

Assuming air to be an ideal gas,

$$h_i - h_e = c_p(T_i - T_e) = \frac{\text{v}_e^2}{2}$$

$$\left(1003.5 \ \frac{\text{J}}{\text{kg·K}}\right)(65°\text{C} - T_e) = \frac{\left(300 \ \frac{\text{m}}{\text{s}}\right)^2}{2}$$

$$T_e = 20.2°\text{C}$$

$$\dot{I} = \dot{m}T_0\Delta s = \dot{m}T_0\left(c_p \ln \frac{T_e}{T_i} - R\ln\frac{P_e}{P_i}\right)$$

$$= \left(1\ \frac{\text{kg}}{\text{s}}\right)(10°\text{C} + 273°)$$

$$\times \left(\begin{array}{l} \left(1.0035\ \frac{\text{kJ}}{\text{kg·K}}\right)\ln\frac{20.2°\text{C} + 273°}{65°\text{C} + 273°} \\ -\left(0.287\ \frac{\text{kJ}}{\text{kg·K}}\right)\ln\frac{100\ \text{kPa}}{180\ \text{kPa}}\end{array}\right)$$

$$= 7.36\ \text{kW}\quad (7.4\ \text{kW})$$

The answer is B.

60. From the psychrometric chart at $T_1 = 45°\text{C}$ and $\phi_1 = 10\%$,

$$\omega_1 = 0.006\ \text{kg water/kg dry air}$$

At $T_{\text{db}} = 33.5°\text{C}$ and $T_{\text{wb}} = 31°\text{C}$,

$$\omega_2 = 0.028\ \text{kg water/kg dry air}$$

The increase in specific humidity is

$$\omega_2 - \omega_1 = 0.028\ \frac{\text{kg water}}{\text{kg dry air}} - 0.006\ \frac{\text{kg water}}{\text{kg dry air}}$$
$$= 0.022\ \text{kg water/kg dry air}$$

The answer is D.

Practice Exam 2

PROBLEMS

1. The net energy change as 20 grams of ethylene polymerizes to polyethylene is

(A) −435 000 J
(B) −348 000 J
(C) −58 600 J
(D) -10×10^{-20} J

2. A metal car bumper is scratched and the scratched area quickly begins to corrode. A child asks why the corrosion happens here and not on the other parts of the bumper. The most accurate answer is

(A) the metal is corroding in many places but this is the only one visible
(B) the scratch in the metal acts as a cathode reaction for corrosion to start
(C) the scratch in the metal acts as an anode reaction for corrosion to start
(D) the metal is locally attacked at the scratch by a corroding environment

3. A metal bar with a 60 mm diameter is pulled in opposite directions on a testing machine with an instantaneous force of 50 kN. The measured elongation of the rod is 0.16 mm in a 500 mm gage length. The diameter of the rod is decreased by 0.01505 mm. Calculate the Young's modulus of the material.

(A) 30×10^9 Pa
(B) 46×10^9 Pa
(C) 55×10^9 Pa
(D) 70×10^9 Pa

4. A motor weighing 50 kg is supported by four springs, each with a spring constant of 2000 N/m. The motor can only move in the vertical direction. The speed at which resonance will occur is

(A) 6 rad/s
(B) 13 rad/s
(C) 40 rad/s
(D) 160 rad/s

5. A homogeneous disk of 5 cm radius and 10 kg mass rotates on an axle AB of length 0.5 m and rotates about a fixed point A. The disk is constrained to roll on a horizontal floor.

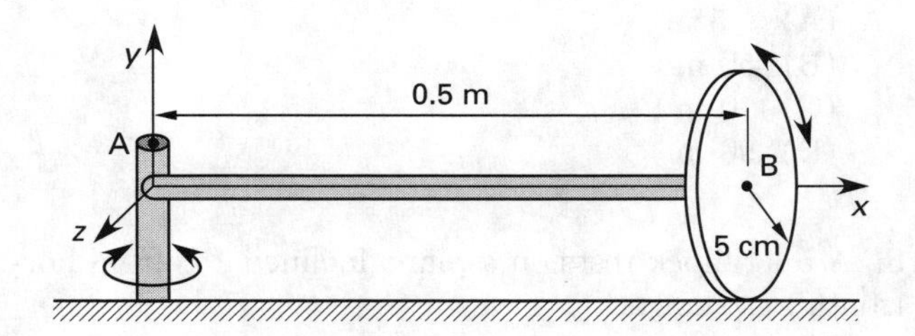

Given an angular velocity of 30 rad/s in the x direction and −3 rad/s in the y direction, the kinetic energy of the disk is

(A) 0.624 J
(B) 16.9 J
(C) 18.0 J
(D) 33.8 J

6. Consider the mass-spring system shown.

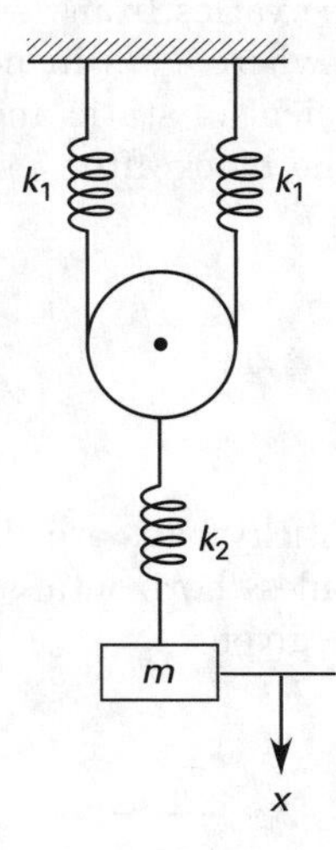

The natural frequency of the system is designed to be $\omega_n = 10$. The spring constant k_2 is half of k_1, and the mass is 1 kg. The mass associated with the other components may be assumed to be negligible. For the given natural frequency, the spring constant k_1 is most nearly

(A) 3 N/m
(B) 40 N/m
(C) 100 N/m
(D) 300 N/m

7. The location of a moving object is given below as a function of time, where x is in meters and t is in seconds.

$$x = t^3 - 9t^2 + 50t - 10$$

The position of the object when its acceleration reaches zero is most nearly

(A) −5 m
(B) 50 m
(C) 70 m
(D) 90 m

8. A 5 kg block rests on a plane inclined 15° from horizontal, as shown.

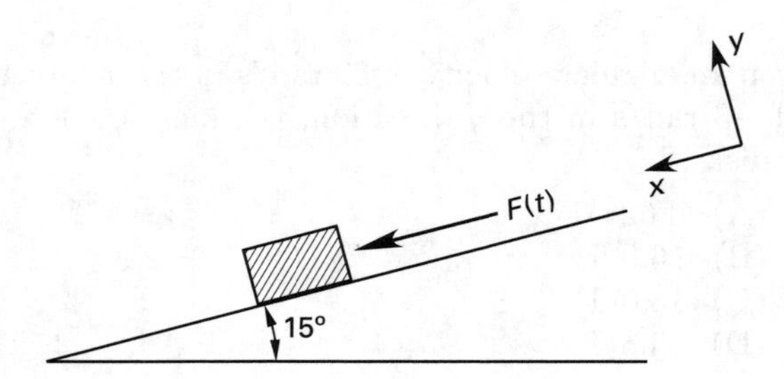

A force parallel to the inclined surface is applied on the block. The force varies from zero according to the function $F = 10t$, where F is in newtons and t is in seconds. The coefficient of static and kinetic friction is 0.3. The speed of the block after 5 s is most nearly

(A) 11 m/s
(B) 13 m/s
(C) 16 m/s
(D) 22 m/s

9. Three identical sticky balls are shown, at an instant of time, on a frictionless horizontal surface. Their associated velocities are given.

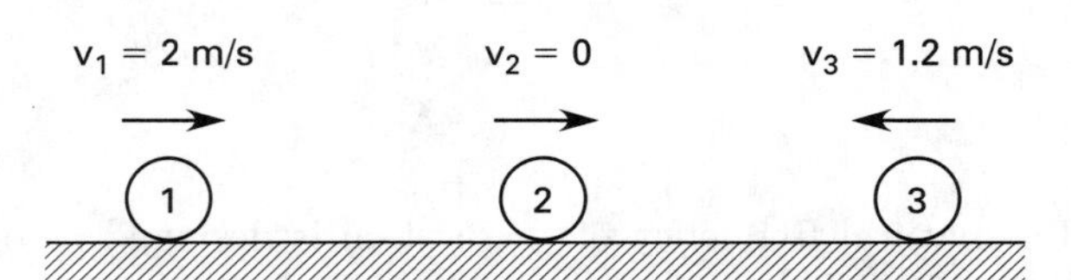

Balls 1 and 2 stick together after impact, and ball 3 bounces off of ball 2. The coefficient of restitution, e, is 1. The velocity of ball 3, after impact, is most nearly

(A) 2 m/s
(B) 3 m/s
(C) 4 m/s
(D) 5 m/s

10. Consider a solid homogeneous disk with uniform thickness, initially at rest on a horizontal surface, as shown.

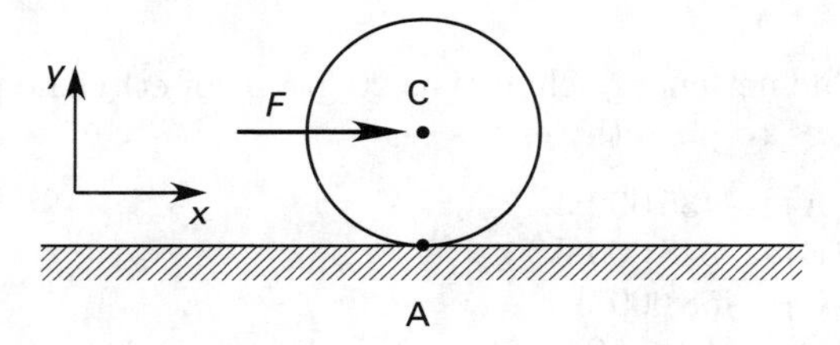

A horizontal force, F, of 350 N is applied to the disk at the center. The coefficients of static and kinetic friction are 0.28 and 0.20, respectively. If the mass and the diameter of the disk are 18 kg and 1.0 m, respectively, the angular acceleration of the disk is most nearly

(A) 5 rad/s^2
(B) 8 rad/s^2
(C) 10 rad/s^2
(D) 20 rad/s^2

11. An air-standard Otto cycle has a compression ratio of 8. At the start of the compression stroke, the temperature is 300K and the pressure is 100 kPa. If the maximum pressure in the cycle is 2.5 MPa the heat input per kg of air is most nearly

(A) 80 kJ/kg
(B) 90 kJ/kg
(C) 100 kJ/kg
(D) 200 kJ/kg

12. An air-standard diesel engine has a compression ratio of 16 and a cutoff ratio of 2. The air enters the engine at 100 kPa and 300K. Assume air to be an ideal gas with constant specific heats ($c_V = 0.7165$ kJ/kg·K, $c_P = 1.0035$ kJ/kg·K). The thermal efficiency of the cycle is most nearly

(A) 30%
(B) 40%
(C) 50%
(D) 60%

13. Air enters the compressor of an air-standard gas turbine at 100 kPa and 300K. The air is compressed to 1 MPa before it enters a combustion chamber, where it is heated to a turbine inlet temperature of 1400K. Assume air to be an ideal gas with constant specific heats. The thermal efficiency is most nearly

(A) 30%
(B) 40%
(C) 50%
(D) 60%

14. A mixture of gases at 30°C and 120 kPa has the following analysis by mass.

$$O_2 = 40\%$$
$$N_2 = 52\%$$
$$H_2 = 8\%$$

The molar mass of the mixture is most nearly

(A) 12 kg/kmol
(B) 14 kg/kmol
(C) 17 kg/kmol
(D) 20 kg/kmol

15. Octane (C_8H_{18}) is burned with 150% theoretical air in a steady combustion process. Assuming complete combustion and a total pressure of 1 atm, the air-fuel ratio is most nearly

(A) 10 kg air/kg fuel
(B) 15 kg air/kg fuel
(C) 20 kg air/kg fuel
(D) 30 kg air/kg fuel

16. A fan is rated to deliver 400 m^3/min of air at a static pressure of 2.5 cm of water when running at 260 rpm and requiring 2.7 kW of power. The efficiency of the fan is most nearly

(A) 40%
(B) 50%
(C) 60%
(D) 70%

17. A fan requires 2.7 kW when running at 260 rpm. If the speed is changed to 300 rpm the power required is most nearly

(A) 3.2 kW
(B) 4.2 kW
(C) 5.2 kW
(D) 5.8 kW

18. The pressure difference across a water pump is 400 kPa. The inlet and outlet areas are equal and the pump is 80% efficient. If the flow rate is 0.015 m^3/s, the power needed is most nearly

(A) 5 kW
(B) 8 kW
(C) 9 kW
(D) 10 kW

19. Oil with a specific gravity of 0.85 is pumped at a volumetric flow rate of 0.015 m^3/s. The inlet and outlet pipes to the pump have the same diameter. The gage pressure is −30 kPa at the pump inlet and is 300 kPa at an elevation of 2 m above the pump outlet. The power required by an ideal pump is most nearly

(A) 2 kW
(B) 3 kW
(C) 4 kW
(D) 5 kW

20. A hovercraft uses an air compressor to provide an air cushion contained within a skirt fastened around the craft. The mass of the craft is 8000 kg, and the craft has a pressurized area of 3 m × 8 m. The average clearance above the water surface is 3 cm. The velocity of the air is low at the center of the cushion. Assuming the air density is 1.2 kg/m^3 and the air is incompressible, the power required by the compressor is most nearly

(A) 110 kW
(B) 120 kW
(C) 160 kW
(D) 250 kW

21. A thin metal disc of mass 0.01 kg is kept balanced by a jet of air, as shown.

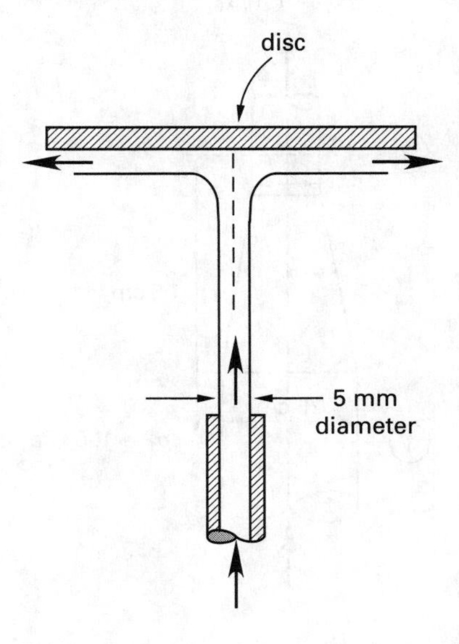

The diameter of the jet at the nozzle exit is 5 mm. Assuming atmospheric conditions at 101.3 kPa and 20°C, the velocity of the jet is most nearly

(A) 45 m/s
(B) 65 m/s
(C) 85 m/s
(D) 95 m/s

22. Water flows in a 5 cm inside-diameter pipe at the rate of 0.004 m^3/s. The length of the pipe is 400 m. If the elevation of the inlet is 20 m above the outlet and the pressure drop is 150 kPa, the friction factor is most nearly

(A) 0.01
(B) 0.02
(C) 0.03
(D) 0.04

23. A submarine is submerged to a depth of 100 m in seawater. The pressure inside the submarine is atmospheric. If the specific gravity of the water is 1.03, the pressure difference across the hull of the submarine is most nearly

(A) 800 kPa
(B) 1000 kPa
(C) 2000 kPa
(D) 3000 kPa

24. Water flows through the reducer shown.

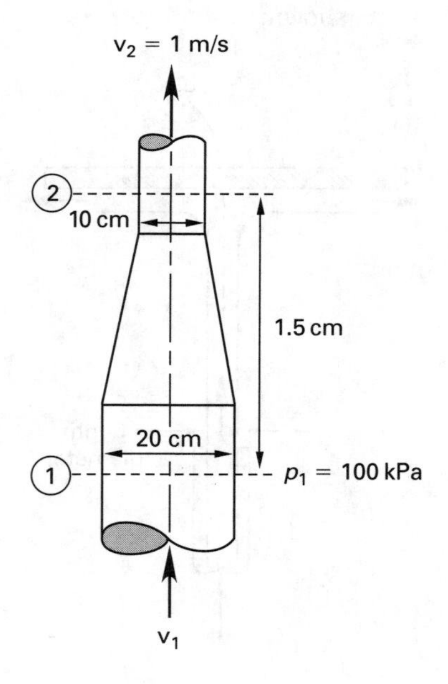

The velocity at section 2 is 1 m/s. Assuming frictionless flow, the pressure at section 2 is most nearly

(A) 40 kPa
(B) 80 kPa
(C) 100 kPa
(D) 200 kPa

25. Oil flows through a 0.12 m diameter pipe at a velocity of 1 m/s. The density and the dynamic viscosity of the oil are 870 kg/m^3 and 0.082 $N\cdot s/m^2$, respectively. If the pipe length is 65 m, the head loss due to friction is most nearly

(A) 1.2 m
(B) 1.5 m
(C) 1.8 m
(D) 2.1 m

26. A window has an area of 1 m $\times$ 1.5 m and a glass thickness of 8 mm. Heat is transferred by conduction through the glass and by convection on the outside and inside surfaces of the window. The convection heat transfer coefficients are $h_i = 1.0\ W/m^2\cdot K$ and $h_o = 35\ W/m^2\cdot K$. The thermal conductivity of the glass is 0.8 $W/m\cdot K$. If the room temperature is maintained at 300K and the outdoor temperature is 270K, the rate of heat loss through the window is most nearly

(A) 220 W
(B) 250 W
(C) 290 W
(D) 320 W

27. A 5 cm diameter copper sphere at a uniform temperature of 1000K is suddenly exposed to a fluid at 300K. The heat transfer coefficient is 400 $W/m^2\cdot K$, and the density and specific heat of copper are 8933 kg/m^3 and 385 $kJ/kg\cdot K$, respectively. Using the lumped capacitance method, the temperature of the sphere after 30 s is most nearly

(A) 700K
(B) 800K
(C) 900K
(D) 1000K

28. 10 kg/s of oil must be cooled from 120°C to 50°C in a concentric counterflow heat exchanger. Water at a temperature of 15°C with a flow rate of 8 kg/s is available. The specific heats of oil and water are 2 $kJ/kg\cdot K$ and 4.18 $kJ/kg\cdot K$, respectively. If the overall heat transfer coefficient is 0.8 $kW/m^2\cdot K$, the heat transfer area is most nearly

(A) 20 m^2
(B) 30 m^2
(C) 40 m^2
(D) 50 m^2

29. Liquid nitrogen at 77K is stored in an uninsulated 1.0 m diameter spherical tank. The tank is vented to the atmosphere and exposed to ambient air at 285K. The latent heat of evaporation for nitrogen is 198 kJ/kg and the density is 810 kg/m^3. Neglect the thermal resistance of the tank and assume the combined convection and radiation heat transfer coefficient is 30 $W/m^2{\cdot}K$. The time it takes for the nitrogen to be completely vented to the atmosphere is most nearly

(A) 0.5 h
(B) 0.8 h
(C) 1 h
(D) 2 h

30. The heat exchange per unit area between two large surfaces at 400K and 300K, if the space between them is evacuated and the surfaces facing each other have an emissivity of 0.02, is most nearly

(A) 4 W/m^2
(B) 6 W/m^2
(C) 8 W/m^2
(D) 10 W/m^2

31. A solid, spherical, mild steel is used as an anchor for a boat. The initial diameter of the steel is 20 cm, and it is exposed to seawater for a period of 4 yr. The steel is losing mass due to corrosion, which is uniform over the surface. The corrosion current developed is estimated to be 6×10^{-5} A/cm^2. The change in diameter of the steel during this period is most nearly

(A) 0.2 cm
(B) 0.6 cm
(C) 0.7 cm
(D) 0.9 cm

32. A single crystal of face-centered cubic (FCC) nickel is subjected to a stress of 16 MPa in the [010] direction of a unit cell. The resolved shear stress on the (111) $[01\bar{1}]$ slip system is most nearly

(A) 6.5 MPa
(B) 7.5 MPa
(C) 8.5 MPa
(D) 9.5 MPa

33. A copolymer consisting of vinyl acetate and vinyl chloride has a molecular mass of 15 000 g/mol and a degree of polymerization (DP) of 200. The mole fraction of vinyl chloride is most nearly

(A) 0.1
(B) 0.3
(C) 0.5
(D) 0.6

34. A 3% C hypoeutectic plain-carbon steel is cooled from 1600°C to 1000°C. The cooling process is under equilibrium conditions.

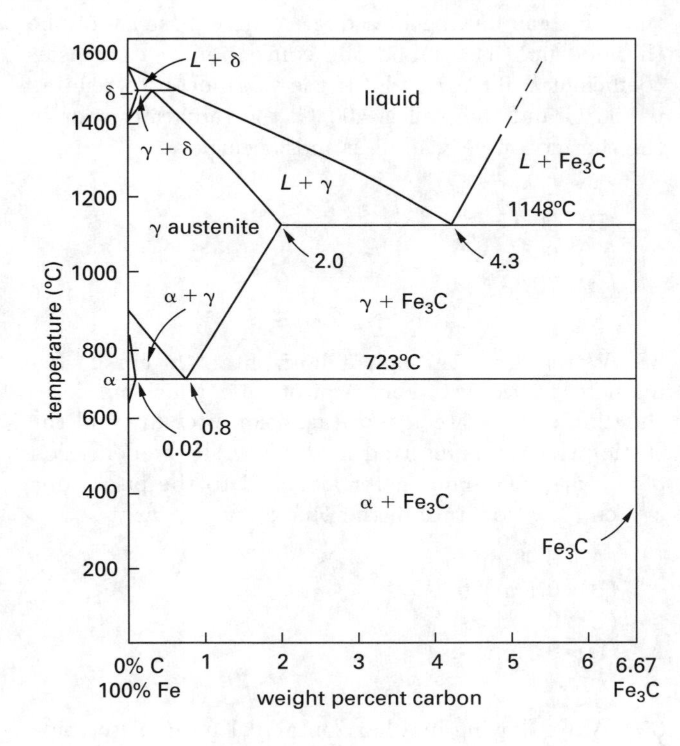

The weight percent of eutectic austenite present in the steel is most nearly

(A) 12%
(B) 17%
(C) 48%
(D) 82%

35. The atomic packing factor (APF) of a body-centered-cubic unit cell (BCC) is most nearly

(A) 0.52
(B) 0.68
(C) 0.72
(D) 0.78

36. To measure low flow rates of air, a laminar flow meter is used. It consists of a large number of small-diameter tubes in parallel. For accurate measurement, the flow in these tubes must be laminar. One design uses 4000 tubes, each with a 2 mm inside diameter and a length of 25 cm. If the pressure difference along the flow meter is 0.5 kPa, the flow rate of atmospheric air at 20°C is most nearly

(A) 0.1 m^3/s
(B) 0.2 m^3/s
(C) 0.4 m^3/s
(D) 0.5 m^3/s

37. A thermocouple is used to measure the temperature of a gas flowing in a duct. The emissivity of the thermocouple is 0.1 and the convection heat transfer coefficient is 90 $W/m^2{\cdot}K$. If the thermocouple reading is 800°C and the wall is 400°C, the radiation error in the thermocouple reading is most nearly

(A) 40°C
(B) 50°C
(C) 60°C
(D) 70°C

38. Water flows through a horizontal, 20 cm inside-diameter frictionless pipe. A pitot tube, which measures the difference between the stagnation pressure and the static pressure, is inserted in the flow. If the deflection of the mercury manometer attached to the pitot tube is 5 cm, the flow rate in the pipe is most nearly

(A) 0.08 m^3/s
(B) 0.1 m^3/s
(C) 0.2 m^3/s
(D) 0.3 m^3/s

39. Water flowing in a horizontal, 0.1 m diameter pipe passes through a venturi meter with a 0.05 m throat diameter. The pressure differential between the inlet and the throat of the venturi meter is 2.5 m of water. If the discharge coefficient is 0.98, the flow rate of the water is most nearly

(A) 0.01 m^3/s
(B) 0.02 m^3/s
(C) 0.03 m^3/s
(D) 0.04 m^3/s

40. A thermocouple is used to measure temperature in a low-temperature application. The generated voltage output of the thermocouple is too low for accurate measurement in this particular application. An amplifier is used to amplify the voltage so that it can be measured conveniently.

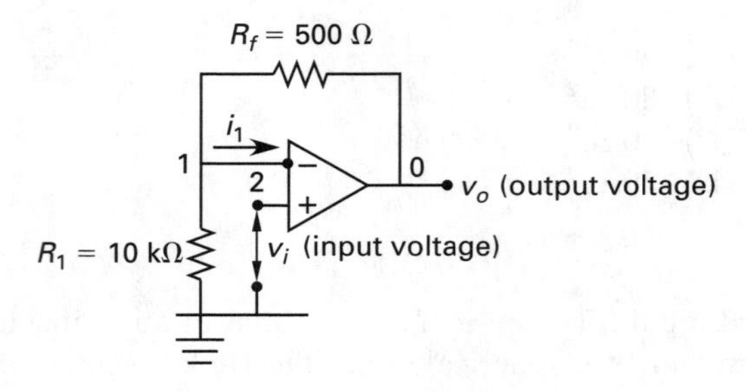

For the circuit shown, the gain of the amplifier is most nearly

(A) 50
(B) 60
(C) 80
(D) 100

41. A wood beam is laminated with a layer of steel plate and mounted as shown.

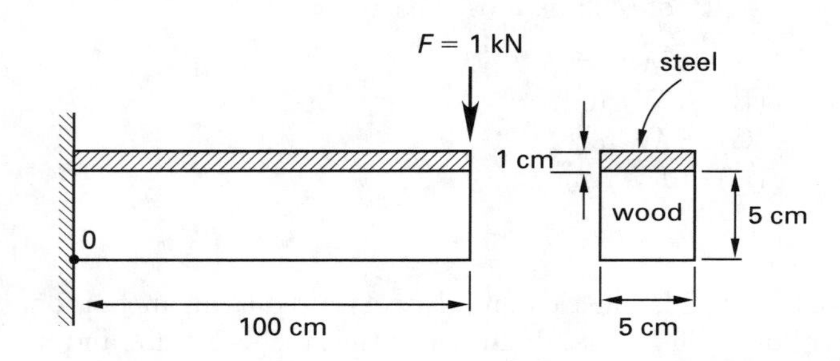

A downward force F is applied at the free end, causing the beam to flex. The moduli of elasticity of the steel and the wood are 195 GPa and 12 GPa, respectively. The maximum bending stress due to force F is most nearly

(A) 40 MPa
(B) 90 MPa
(C) 200 MPa
(D) 300 MPa

42. A 100 kg disk is mounted on a solid shaft of diameter D, as shown. The system is designed to rotate near its critical speed of 500 rpm. The modulus of elasticity of the shaft is 200 GPa.

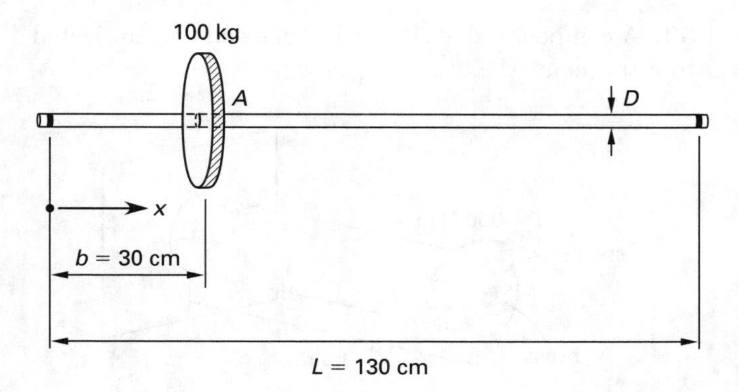

For the configuration shown, the shaft diameter is most nearly

(A) 4 mm
(B) 10 mm
(C) 20 mm
(D) 30 mm

43. A 1 m long solid steel shaft 30 mm in diameter transmits 35 kW of power. The allowable shear stress is 70 MPa and the shear modulus is 75 GPa. The total angular twist is not to exceed 1.5°. The maximum angular speed should most nearly be

(A) 400 rpm
(B) 900 rpm
(C) 1000 rpm
(D) 2000 rpm

44. A cylindrical pressure vessel is constructed to operate at an internal gage pressure of 1.5 MPa. The vessel is to carry a longitudinal force, F, as shown.

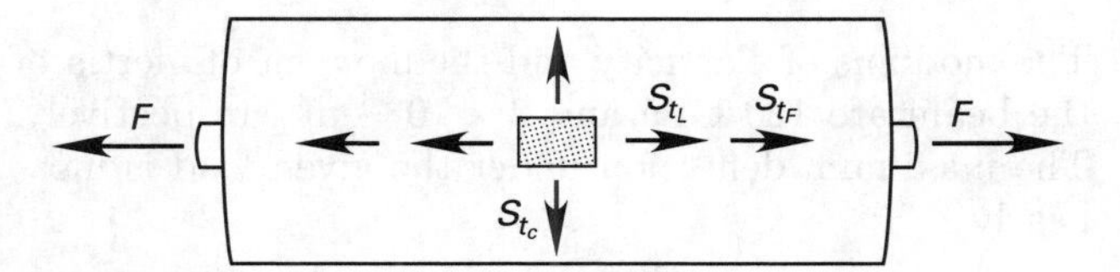

The vessel wall thickness and inside diameter are 1 cm and 70 cm, respectively. The allowable tensile stress of the steel is 105 MPa. The maximum longitudinal force that the vessel wall can handle is most nearly

(A) 1.7 MN
(B) 2.7 MN
(C) 3.7 MN
(D) 4.7 MN

45. A round concrete bar 10 cm long is struck with a 20 kg object moving horizontally with a velocity of 20 m/s. The allowable compressive stress in the bar is 10 MPa and its modulus is 17 MPa.

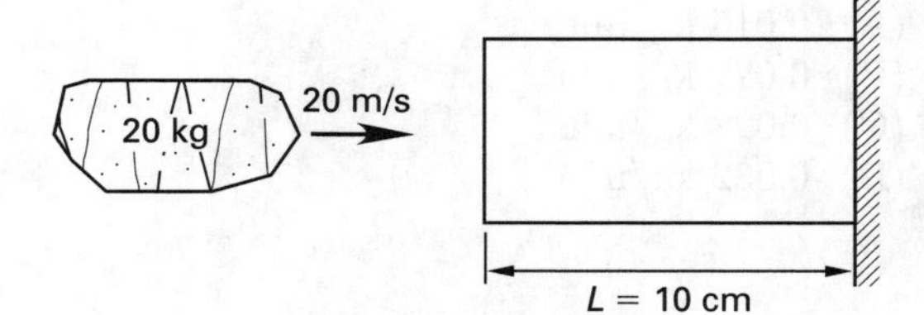

Assume that the mass of the moving object is very large compared to the mass of the bar and, after impact, all of the energy is transformed into strain energy. The minimum required diameter of the bar is most nearly

(A) 5.8 cm
(B) 9.5 cm
(C) 13 cm
(D) 17 cm

46. An ideal vapor-compression refrigeration cycle using refrigerant 134a, operates between an evaporator pressure of 0.12 MPa and a condenser pressure of 0.8 MPa. The coefficient of performance of the cycle is most nearly

(A) 2.6
(B) 3.6
(C) 4.6
(D) 5.6

47. A room contains air at 25°C and 101.3 kPa at a relative humidity of 70%. The dew point temperature is most nearly

(A) 11°C
(B) 13°C
(C) 16°C
(D) 19°C

48. Air enters an air-conditioning unit at 1 atm, 28°C, and 80% relative humidity at the rate of 15 m^3/min. The air is dehumidified and leaves as saturated air at 12°C. If the condensed moisture is also at 12°C, the rate of heat transfer is most nearly

(A) −800 kJ/min
(B) −700 kJ/min
(C) −600 kJ/min
(D) −400 kJ/min

49. Atmospheric air at a temperature of 20°C and a relative humidity of 80% flows in a duct at the rate of 2 m^3/min. If the air is cooled to 5°C, using the psychrometric charts, the rate of condensation of the water vapor is most nearly

(A) 0.015 kg/min
(B) 0.021 kg/min
(C) 0.028 kg/min
(D) 0.032 kg/min

50. 5 kg of an air-water vapor mixture at 28°C and 60% relative humidity are cooled at a constant pressure of 101.3 kPa to 5°C.

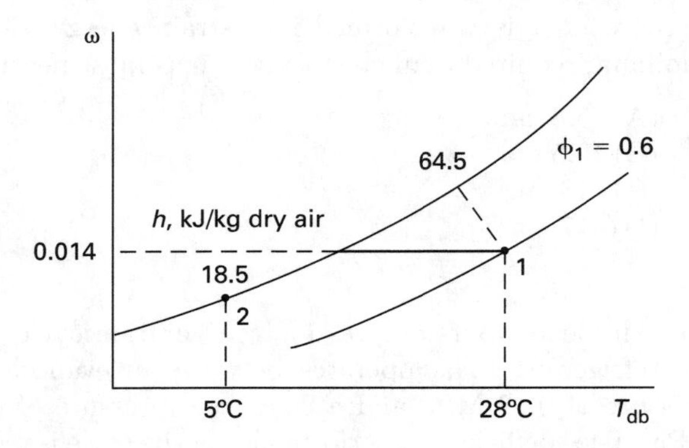

The heat transfer is most nearly

(A) −240 kJ
(B) −230 kJ
(C) −210 kJ
(D) −180 kJ

51. The steel structure shown is subject to a 5 kN downward force at point C as shown. Members AC and BC have a length and cross-sectional area of 5 m and 1 cm^2, respectively. The modulus of elasticity and allowable stress are 200 GPa and 200 MPa, respectively.

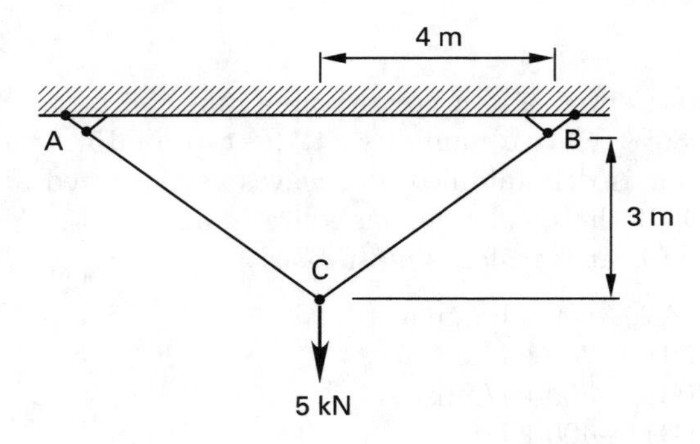

The stress in member AC is most nearly

(A) 15 MPa
(B) 25 MPa
(C) 41 MPa
(D) 55 MPa

52. A composite shaft is fixed at one end and subjected to a torque at the other, as shown.

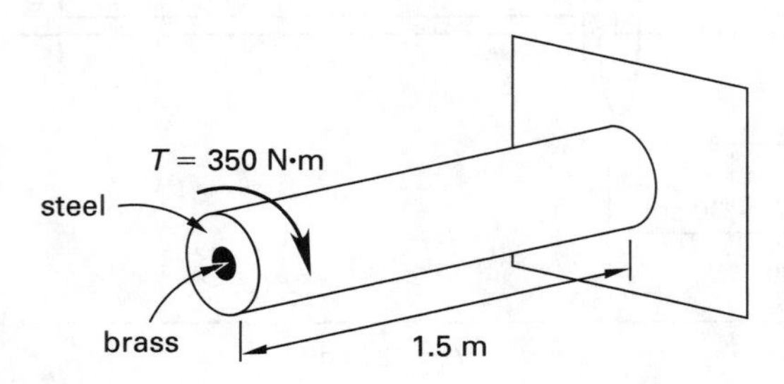

The shaft consists of a solid brass rod inside a steel tube. The radius of the brass rod and the outside radius of the steel tube are 10 mm and 30 mm, respectively. The length of the shaft is 1.5 m. The shear moduli of brass and of steel are 35 GPa and 75 GPa, respectively. The angle of rotation at the free end of the shaft due to the 350 N·m torque is most nearly

(A) 0.1°
(B) 0.3°
(C) 0.9°
(D) 2°

53. A beam of uniform cross section is loaded as shown.

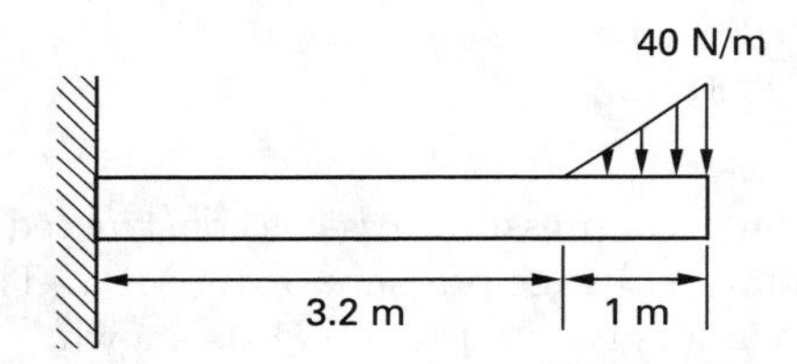

The modulus of elasticity and the moment of inertia of the beam are 120 GPa and 2×10^{-6} m^4, respectively. The maximum deflection under the given load is most nearly

(A) 2 mm
(B) 5 mm
(C) 9 mm
(D) 20 mm

54. A copper pipe is fitted between two walls, as shown.

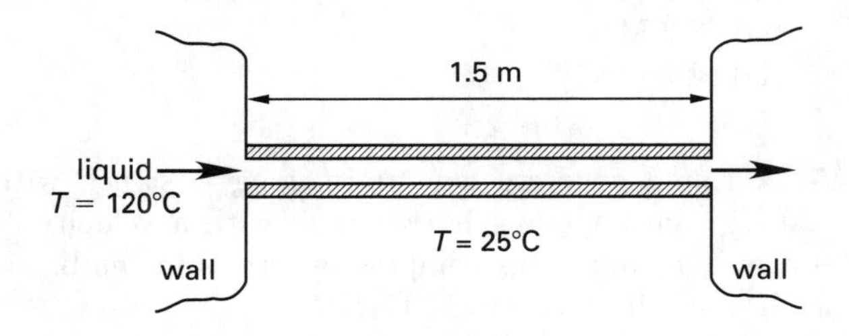

A hot liquid flows through the pipe at an average temperature of 120°C, and the surrounding temperature is 25°C. The inside and outside pipe diameters are 2.4 cm and 2.6 cm, respectively. The pipe length is 1.5 m. The pipe modulus is 100 GPa, and its coefficient of thermal expansion is 19×10^{-6}/°C. The average stress applied at the wall is most nearly

(A) 150 MPa
(B) 160 MPa
(C) 170 MPa
(D) 180 MPa

55. A cantilever I-beam is subjected to a force of 4 kN, as shown.

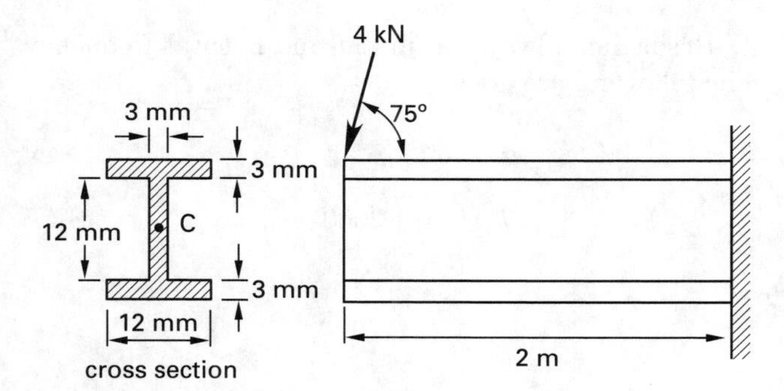

The maximum tensile stress is most nearly

(A) 11 GPa
(B) 12 GPa
(C) 13 GPa
(D) 14 GPa

56. Air enters the compressor of an ideal air-standard gas turbine at 100 kPa and 300K. The pressure ratio is 10 and the turbine inlet temperature is 1300K. Assuming a constant-pressure specific heat, $c_p = 1.087$ kJ/kg·K, and $k = c_p/c_V = 1.36$, the thermal efficiency is most nearly

(A) 31%
(B) 35%
(C) 41%
(D) 46%

57. 1 kg of saturated steam at 200 kPa is compressed reversibly and adiabatically (isentropically) to 2 MPa. It then undergoes a constant-pressure cooling until it reaches a quality of 0.3. The total heat interaction is most nearly

(A) −1800 kJ/kg
(B) −1500 kJ/kg
(C) −1200 kJ/kg
(D) −1100 kJ/kg

58. Steam at 4 MPa and 400°C enters an adiabatic turbine operating at steady state. The steam leaves the turbine at 100 kPa and 150°C. If the mass flow rate through the turbine is 3 kg/s, the isentropic efficiency of the turbine is most nearly

(A) 30%
(B) 40%
(C) 60%
(D) 70%

59. The highest and lowest pressures of a Carnot power cycle utilizing air as a working fluid are 600 kPa and 100 kPa. The lowest temperature in the cycle is 300K, and the pressures at the end of the heat transfer processes are identical.

Assuming air to be an ideal gas with constant specific heats, ($c_p = 1.0035$ kJ/kg·K, $c_V = 0.7165$ kJ/kg·K, and $R = 0.287$ kJ/kg·K), the work developed during the cycle is most nearly

(A) 12 kJ/kg
(B) 22 kJ/kg
(C) 32 kJ/kg
(D) 42 kJ/kg

60. Air flows through an air compressor in steady-state operation. The temperature and pressure at the inlet are 300K and 100 kPa, respectively, while the exit temperature and pressure are 500K and 900 kPa, respectively. The compression process may be modeled as a polytropic compression ($pv^n = C$). Assuming air to be an ideal gas with constant specific heats, the work transfer per unit mass of air is most nearly

(A) 210 kJ/kg
(B) 250 kJ/kg
(C) 310 kJ/kg
(D) 350 kJ/kg

SOLUTIONS

1. Each added C_2H_4 molecule breaks one C=C bond and forms two C-C bonds. C=C bond energy is 614 kJ/mol, which means 614 kJ/mol of energy are required to break 0.602×10^{24} C=C bonds. C-C bond energy is 348 kJ/mol, which means 348 kJ/mol of energy are required to break 0.602×10^{24} C-C bonds.

The net bond energy for this process is

$$\begin{aligned}&\frac{614\,000 \text{ J}}{0.602 \times 10^{24} \text{ molecules}} - \frac{(2)(348\,000 \text{ J})}{0.602 \times 10^{24} \text{ molecules}}\\ &\quad = -1.36 \times 10^{-19} \frac{\text{J}}{\text{C}_2\text{H}_4}\end{aligned}$$

For 20 grams of C_2H_4,

$$\begin{aligned}&\frac{m_{\text{grams}}\text{mol}_{\text{amu/gram}}}{\text{MW}_{\text{C}_2\text{H}_4}}\\ &\quad = \frac{(20 \text{ g})\left(0.602 \times 10^{24} \dfrac{\text{amu}}{\text{g}}\right)}{28 \dfrac{\text{amu}}{\text{C}_2\text{H}_4}}\\ &\quad = 4.3 \times 10^{23} \text{ molecules of C}_2\text{H}_4\end{aligned}$$

The net energy change is

$$\begin{aligned}E_{\text{net}} &= (\text{net bond energy})(\text{molecules of C}_2\text{H}_4)\\ &= \left(-1.36 \times 10^{-19} \frac{\text{J}}{\text{C}_2\text{H}_4}\right)(4.3 \times 10^{23} \text{ C}_2\text{H}_4)\\ &= -58\,566 \text{ J} \quad (-58\,600 \text{ J})\end{aligned}$$

The answer is C.

2. Car bumpers are often made of a corrodible metal that's covered with a thin plating, made mostly of chromium (known as chrome), which makes them shiny and forms a protective coating that oxygen can't penetrate. When the chrome is scratched, the metal under the chrome is exposed to the air. The metal then begins to oxidize, which causes corrosion. Oxidation is an anode reaction.

The answer is C.

3. Young's modulus is given by the equation

$$E = \frac{PL_o}{A\delta}$$

The area of the 60 mm bar is

$$\begin{aligned}A &= \tfrac{1}{4}\pi d_o^2 = \left(\tfrac{1}{4}\pi\right)(60 \text{ mm})\left(\frac{1 \text{ m}}{1000 \text{ mm}}\right)\\ &= 2.826 \times 10^{-3} \text{ m}^2\end{aligned}$$

Plugging this value into Young's modulus,

$$\begin{aligned}E &= \frac{(50\,000 \text{ N})(0.5 \text{ m})}{(2.826 \times 10^{-3} \text{ m}^2)(1.6 \times 10^{-4} \text{ m})}\\ &= 55.30 \times 10^9 \text{ Pa} \quad (55 \times 10^9 \text{ Pa})\end{aligned}$$

The answer is C.

4. Resonance always occurs at the natural frequency. The following are given.

$$\begin{aligned}m &= 50 \ kg\\ k &= (4)\left(2000 \frac{\text{N}}{\text{m}}\right)\\ &= 8000 \frac{\text{N}}{\text{m}}\end{aligned}$$

The natural frequency is

$$\begin{aligned}\omega &= \sqrt{\frac{k}{m}} = \sqrt{\frac{8000 \dfrac{\text{N}}{\text{m}}}{50 \text{ kg}}}\\ &= 12.65 \text{ rad/s} \quad (13 \text{ rad/s})\end{aligned}$$

The answer is B.

5. Assuming the axle is part of the disk, the disk has a fixed point at A. Since the x, y, and z axes are principal axes of inertia for the disk, the kinetic energy is

$$\begin{aligned}\text{KE} &= \tfrac{1}{2}\left(I_x\omega_x^2 + I_y\omega_y^2 + I_z\omega_z^2\right)\\ &= \tfrac{1}{2}\left(\tfrac{1}{2}mr^2\omega_x^2 + \left(mL^2 + \tfrac{1}{4}mr^2\right)\omega_y^2 + 0\right)\\ &= \left(\frac{1}{2}\right)\left(\begin{array}{c}\left(\dfrac{1}{2}\right)(10 \text{ kg})(0.05 \text{ m})^2\left(30 \dfrac{\text{rad}}{\text{s}}\right)^2\\ + \left(\begin{array}{c}(10 \text{ kg})(0.5 \text{ m})^2\\ + \left(\dfrac{1}{4}\right)(10 \text{ kg})(0.05 \text{ m})^2\end{array}\right)\\ \times \left(-3 \dfrac{\text{rad}}{\text{s}}\right)^2\end{array}\right)\\ &= 16.9 \text{ J}\end{aligned}$$

The answer is B.

6. The equivalent system is represented by

$$\ddot{x} + \left(\frac{k}{m}\right) x = 0$$

$$\omega = \sqrt{\frac{k}{m}} \qquad \textit{Eq. 1}$$

$$k = \left(\frac{1}{2k_1} + \frac{1}{k_2}\right)^{-1} = \left(\frac{k_2 + 2k_1}{2k_1 k_2}\right)^{-1} \qquad \textit{Eq. 2}$$

For $k_2 = {}^1\!/_2 k_1$, $m = 1$ kg, $\omega = 10$, $k = {}^2\!/_5 k_1$, using Eqs. 1 and 2,

$$10 = \sqrt{\frac{\frac{2}{5}k_1}{1}}$$

$$k_1 = 250 \text{ N/m} \quad (300 \text{ N/m})$$

The answer is D.

7. Velocity is obtained by differentiating the position function.

$$\text{v} = \frac{dx}{dt} = 3t^2 - 18t + 50$$

Acceleration is obtained by differentiating the velocity function.

$$a = \frac{d\text{v}}{dt} = 6t - 18$$

Solve for time when acceleration is zero.

$$0 = 6t - 18$$
$$t = 3 \text{ s}$$

Substitute $t = 3$ s into the position equation to obtain the location of the object, x, in meters.

$$\begin{aligned} x &= t^3 - 9t^2 + 50t - 10 \\ &= (3)^3 - (9)(3)^2 + (50)(3) - 10 \\ &= 86 \text{ m} \quad (90 \text{ m}) \end{aligned}$$

The answer is D.

8. The impulse-momentum method applies. The free-body diagram is as follows.

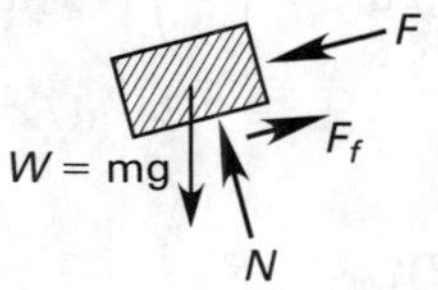

Step 1: Find the time when the block starts to slide.

The force needed to get the block to slide is calculated as

$$\sum F_x = 0$$

$$F + W \sin 15° - F_f = 0 \qquad \textit{Eq. 1}$$

$$W = mg$$

$$F_f = \mu N \qquad \textit{Eq. 2}$$

$$F = \mu N - mg \sin 15°$$

$$\sum F_y = 0$$

$$N - W \cos 15° = 0 \qquad \textit{Eq. 3}$$

$$N = mg \cos 15°$$

From Eqs. 1, 2, and 3,

$$\begin{aligned} F &= mg(\mu \cos 15° - \sin 15°) \\ &= (5 \text{ kg})\left(9.81 \ \frac{\text{m}}{\text{s}^2}\right)((0.3)(0.966) - 0.259) \\ &= 1.571 \text{ N} \end{aligned}$$

Since $F = 10t$,

$$\begin{aligned} t &= \frac{F}{10} = \frac{1.571 \text{ N}}{10} \\ &= 0.151 \text{ s} \end{aligned}$$

Step 2:

The impulse-momentum equation is

$$\int_{t_1}^{t_2} F_e(t)\, dt = m\left(\text{v}(t_2) - \text{v}(t_1)\right)$$

F_e is the external forces on the block.

$$\int_{t_1}^{t_2} F\, dt - \int_{t_1}^{t_2} F_f\, dt = m(\text{v}_f - \text{v}_i) \qquad \textit{Eq. 4}$$

$$F = 10t$$

$$\begin{aligned} F_f &= \mu N = \mu(W \cos 15°) \\ &= \mu mg \cos 15° \\ &= (0.3)(5 \text{ kg})\left(9.81 \ \frac{\text{m}}{\text{s}^2}\right) \\ &\quad \times \cos 15° \\ &= 14.2 \text{ N} \end{aligned}$$

Using Eq. 4,

$$\int_{0.151 \text{ s}}^{5 \text{ s}} 10t\, dt - 14.2 \int_{0.151 \text{ s}}^{5 \text{ s}} dt = (5 \text{ kg})(\text{v}_f - 0)$$

$$\text{v}_f = 11.2 \text{ m/s} \quad (11 \text{ m/s})$$

The answer is A.

9. Apply the law of conservation of linear momentum.

$$(m\mathrm{v})_{\text{ball 1}} + (m\mathrm{v})_{\text{ball 2}} = (2m\mathrm{v})_{\text{combined 1 \& 2}}$$

Since $m_1 = m_2$, the combined velocity of balls 1 and 2 after impact is

$$\begin{aligned} \mathrm{v}_c \equiv \mathrm{v}_{\text{combined}} &= \tfrac{1}{2}(\mathrm{v}_1 + \mathrm{v}_2) \\ &= \left(\frac{1}{2}\right)\left(2\ \frac{\mathrm{m}}{\mathrm{s}} + 0\ \frac{\mathrm{m}}{\mathrm{s}}\right) \\ \mathrm{v}_c &= 1\ \mathrm{m/s}\ \text{to right} \end{aligned}$$

Apply the principle of conservation of linear momentum again between the combined mass of balls 1, 2, and 3.

$$2m\mathrm{v}_c + m_3\mathrm{v}_3 = 2m\mathrm{v}_c' + m\mathrm{v}_3' \qquad \textit{Eq. 1}$$

v_c' and v_3' are velocities after impact.

Simplify Eq. 1.

$$\begin{aligned} (2)\left(1\ \frac{\mathrm{m}}{\mathrm{s}}\right) + \left(-1.2\ \frac{\mathrm{m}}{\mathrm{s}}\right) &= 2\mathrm{v}_c' + \mathrm{v}_3' \\ 2\mathrm{v}_c' + \mathrm{v}_3' &= 0.8\ \mathrm{m/s} \qquad \textit{Eq. 2} \end{aligned}$$

Using the coefficient of restitution, e,

$$\mathrm{v}_3' - \mathrm{v}_c' = e(\mathrm{v}_c - \mathrm{v}_3)$$

Since $e = 1$,

$$\begin{aligned} \mathrm{v}_3' - \mathrm{v}_c' &= (1)\left(1\ \frac{\mathrm{m}}{\mathrm{s}} - \left(-1.2\ \frac{\mathrm{m}}{\mathrm{s}}\right)\right) \\ \mathrm{v}_3' - \mathrm{v}_c' &= 2.2\ \mathrm{m/s} \qquad \textit{Eq. 3} \end{aligned}$$

Using Eqs. 2 and 3,

$$\begin{aligned} \mathrm{v}_c' &= -0.47\ \mathrm{m/s} \\ \mathrm{v}_3' &= 1.73\ \mathrm{m/s} \quad (2\ \mathrm{m/s}) \end{aligned}$$

The answer is A.

10. Apply the law of conservation of forces at the center of gravity.

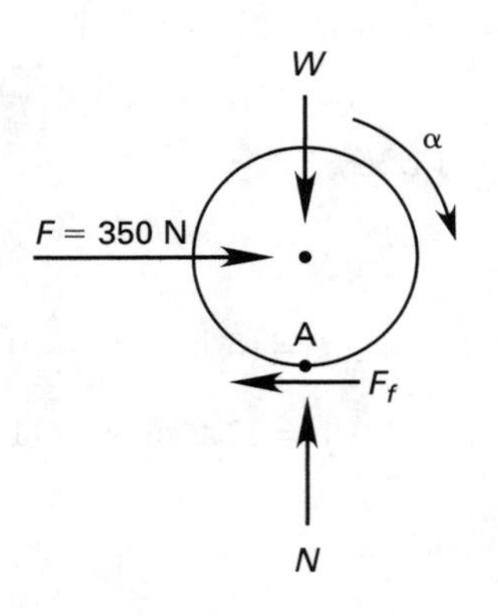

$$\begin{aligned} \sum F_{x_c} &= ma_{c_x} \\ 350N - F_f &= mr\alpha \\ &= (18\ \mathrm{kg})(0.5\ \mathrm{m})\alpha \qquad \textit{Eq. 1} \end{aligned}$$

$$\begin{aligned} \sum F_{y_c} &= ma_{c_y} = 0 \\ N - W &= 0 \\ N = W &= mg \\ &= (18\ \mathrm{kg})\left(9.81\ \frac{\mathrm{m}}{\mathrm{s}^2}\right) \\ &= 176.6\ \mathrm{N} \end{aligned}$$

Assume no slippage and apply the law of conservation of momentum at A.

$$\begin{aligned} \sum M_A &= Fr = I_A\alpha \\ (350\ \mathrm{N})(0.5\ \mathrm{m}) &= \left(\tfrac{1}{2}mr^2 + mr^2\right)\alpha \\ 175\ \mathrm{N{\cdot}m} &= \left(\begin{array}{c}\left(\tfrac{1}{2}\right)(18\ \mathrm{kg})(0.5\ \mathrm{m})^2 \\ +\ (18\ \mathrm{kg})(0.5\ \mathrm{m})^2\end{array}\right)\alpha \\ \alpha &= 25.9\ \mathrm{rad/s^2} \end{aligned}$$

Using Eq. 1,

$$F_f = 116.9\ \mathrm{N}$$

Check for slippage.

$$\begin{aligned} F_{f_{\max}} &= \mu_s N = (0.28)(176.6\ \mathrm{N}) \\ &= 49.5\ \mathrm{N} \end{aligned}$$

Since $F_{f_{\max}} < F_f$, slippage does occur.

$$\begin{aligned} \sum F_{x_c} &= F - F_f \\ &= F - \mu_k N \\ &= ma_{x_c} \\ 350\ \mathrm{N} - (0.2)(176.6\ \mathrm{N}) &= (18\ \mathrm{kg})a_{x_c} \\ a_{x_c} &= 17.5\ \mathrm{m/s^2} \end{aligned}$$

$$\begin{aligned} \sum M_c &= F_f r = I_c\alpha \qquad \textit{Eq. 2} \\ F_f &= \mu_k N = (0.2)(176.6\ \mathrm{N}) \\ &= 35.3\ \mathrm{N} \end{aligned}$$

Using Eq. 2,

$$\begin{aligned} F_f r &= \tfrac{1}{2}mr^2\alpha \\ (35.3\ \mathrm{N})(0.5\ \mathrm{m}) &= \left(\frac{1}{2}\right)(18\ \mathrm{kg})(0.5\ \mathrm{m})^2\alpha \\ \alpha &= 7.8\ \mathrm{rad/s^2} \quad (8\ \mathrm{rad/s^2}) \end{aligned}$$

The answer is B.

11.

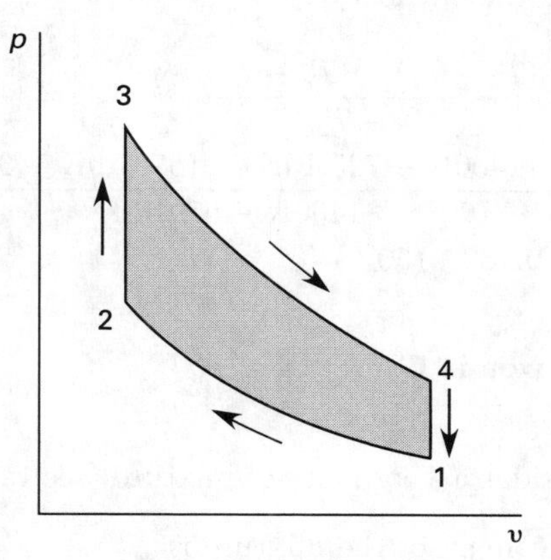

Otto cycle

$$\frac{T_2}{T_1} = \left(\frac{p_2}{p_1}\right)^{\frac{k-1}{k}} = \left(\frac{v_1}{v_2}\right)^{k-1}$$
$$= r_v^{k-1}$$
$$T_2 = T_1(r_v)^{k-1}$$
$$= (300\text{K})(8)^{1.4-1}$$
$$= 689.22\text{K}$$
$$v_1 = \frac{RT_1}{p_1}$$
$$= \frac{\left(0.287\ \frac{\text{kJ}}{\text{kg·K}}\right)(300\text{K})}{100\ \text{kPa}}$$
$$= 0.861\ \text{m}^3/\text{kg}$$
$$v_3 = v_2 = \frac{v_1}{r_v}$$
$$= \frac{0.861\ \frac{\text{m}^3}{\text{kg}}}{8}$$
$$= 0.1076\ \text{m}^3/\text{kg}$$
$$T_3 = \frac{p_3 v_3}{R}$$
$$= \frac{(2500\ \text{kPa})\left(0.1076\ \frac{\text{m}^3}{\text{kg}}\right)}{0.287\ \frac{\text{kJ}}{\text{kg·K}}}$$
$$= 937.5\text{K}$$

The heat input is

$$q_{2\text{-}3} = u_3 - u_2 = c_V(T_3 - T_2)$$
$$= \left(0.7165\ \frac{\text{kJ}}{\text{kg·K}}\right)(937.5\text{K} - 689.22\text{K})$$
$$= 177.89\ \text{kJ/kg} \quad (200\ \text{kJ/kg})$$

The answer is D.

12.

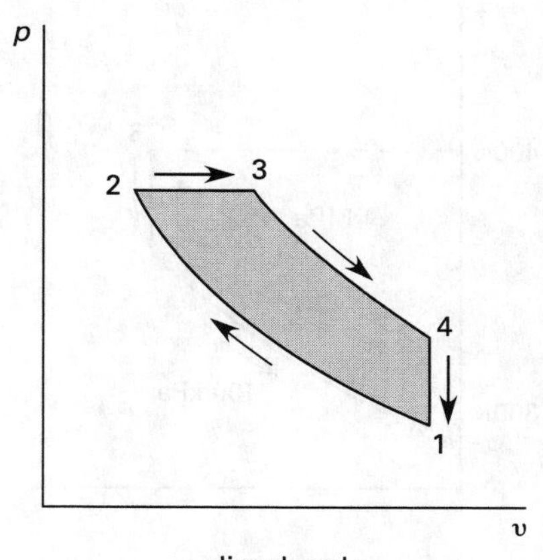

diesel cycle

The efficiency of the cycle is

$$\eta = 1 - \frac{u_4 - u_1}{h_3 - h_2}$$
$$= 1 - \frac{T_4 - T_1}{k(T_3 - T_2)}$$

Processes 1-2 and 3-4 are isentropic so that

$$\frac{T_2}{T_1} = \left(\frac{v_1}{v_2}\right)^{k-1} = (r_v)^{k-1}$$
$$T_2 = T_1(r_v)^{k-1} = (300\text{K})(16)^{1.4-1}$$
$$= 909.43\text{K}$$

Process 2-3 is at constant pressure so that

$$\frac{T_3}{T_2} = \frac{v_3}{v_2} = r_c$$
$$T_3 = T_2 r_c = (909.43\text{K})(2)$$
$$= 1818.86\text{K}$$
$$\frac{T_4}{T_3} = \left(\frac{v_3}{v_4}\right)^{k-1} = \left(\left(\frac{v_3}{v_2}\right)\left(\frac{v_2}{v_4}\right)\right)^{k-1}$$
$$= \left((2)\left(\frac{1}{16}\right)\right)^{1.4-1}$$
$$= 0.43528$$
$$T_4 = (1818.86\text{K})(0.43528)$$
$$= 791.7\text{K}$$

The efficiency is

$$\eta = 1 - \frac{T_4 - T_1}{k(T_3 - T_2)}$$
$$= 1 - \frac{791.7\text{K} - 300\text{K}}{(1.4)(1818.86\text{K} - 909.43\text{K})}$$
$$= 61.38\% \quad (60\%)$$

The answer is D.

13.

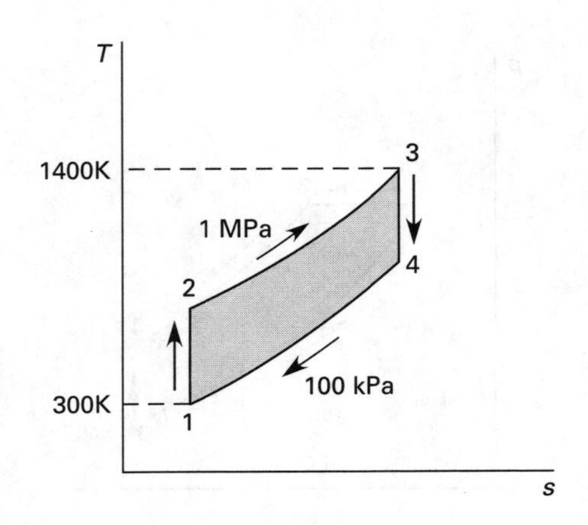

The efficiency is given by

$$\eta = \frac{(h_3 - h_4) - (h_2 - h_1)}{h_3 - h_2} = \frac{(T_3 - T_4) - (T_2 - T_1)}{T_3 - T_2}$$

The pressure ratio is

$$r_p = \frac{p_{23}}{p_{41}} = \frac{(1\ \text{MPa})\left(1000\ \frac{\text{kPa}}{\text{MPa}}\right)}{100\ \text{kPa}} = 10$$

From isentropic relations,

$$\begin{aligned} T_2 &= T_1 (r_p)^{\frac{k-1}{k}} \\ &= (300\text{K})10^{\frac{1.4-1}{1.4}} \\ &= (300\text{K})(1.9306) \\ &= 579.19\text{K} \end{aligned}$$

$$\begin{aligned} T_4 &= T_3 \left(\frac{1}{r_p}\right)^{\frac{k-1}{k}} \\ &= (1400\text{K})\left(\frac{1}{10}\right)^{\frac{1.4-1}{1.4}} \\ &= (1400\text{K})(0.5179) \\ &= 725.15\text{K} \end{aligned}$$

Therefore the efficiency is

$$\begin{aligned} \eta &= \frac{(T_3 - T_4) - (T_2 - T_1)}{T_3 - T_2} \\ &= \frac{(1400\text{K} - 725.15\text{K}) - (579.19\text{K} - 300\text{K})}{1400\text{K} - 579.19\text{K}} \\ &= 0.482 \quad (50\%) \end{aligned}$$

The answer is C.

14. Consider a 1 kg mass of mixture (see table).

The molar mass of the mixture is

$$\begin{aligned} \text{MW} &= \sum x_i (\text{MW})_i \\ &= (0.1766)\left(32\ \frac{\text{kg}}{\text{kmol}}\right) \\ &\quad + (0.2627)\left(28.013\ \frac{\text{kg}}{\text{kmol}}\right) \\ &\quad + (0.5607)\left(2.016\ \frac{\text{kg}}{\text{kmol}}\right) \\ &= 14.14\ \text{kg/kmol} \quad (14\ \text{kg/kmol}) \end{aligned}$$

The answer is B.

15. The stoichiometric equation is

$$\begin{aligned} &C_8H_{18} + 12.5(O_2 + 3.76N_2) \\ &\quad \rightarrow 8CO_2 + 9H_2O + (12.5)(3.76)N_2 \end{aligned}$$

With 150% theoretical air,

$$\begin{aligned} &C_8H_{18} + (1.5)(12.5)(O_2 + 3.76)N_2 \\ &\quad \rightarrow 8CO_2 + 9H_2O + (1.5)(12.5)(3.76)N_2 \\ &\quad + (0.5)(12.5)O_2 \end{aligned}$$

Alternatively,

$$\begin{aligned} &C_8H_{18} + 18.75O_2 + 70.5N_2 \\ &\quad \rightarrow 8CO_2 + 9H_2O + 70.5N_2 + 6.25O_2 \end{aligned}$$

Table for Sol. 14

	mass fraction, m_i	$(\text{MW})_i$ (kg/kmol)	$n_i = \frac{m_i}{(\text{MW})_i}$	mol fraction, $x_i = \frac{n_i}{\sum n_i}$
O_2	0.4	32	0.0125	0.1766
N_2	0.52	28.013	0.0186	0.2627
H_2	0.08	2.016	0.0397	0.5607
	1.00		$\sum n_i = 0.0708$	1.000

The air-fuel ratio is

$$\frac{\text{A}}{\text{F}} = \frac{m_{\text{air}}}{m_{\text{fuel}}} = \frac{\left(\begin{array}{l} (18.75 \text{ kmol}) \left(32 \, \frac{\text{kg}}{\text{kmol}}\right) \\ +(70.5 \text{ kmol}) \left(28.013 \, \frac{\text{kg}}{\text{kmol}}\right) \end{array} \right)}{\left(\begin{array}{l} (8 \text{ kmol}) \left(12.01 \, \frac{\text{kg}}{\text{kmol}}\right) \\ +(9 \text{ kmol}) \left(2.016 \, \frac{\text{kg}}{\text{kmol}}\right) \end{array} \right)}$$

$$= 22.54 \text{ kg air/kg fuel} \quad (20 \text{ kg air/kg fuel})$$

The answer is C.

16. Ideal power is given by

$$\begin{aligned} P_{\text{ideal}} &= Q\Delta p \\ &= \left(400 \, \frac{\text{m}^3}{\text{min}}\right) \left(\frac{1 \text{ min}}{60 \text{ s}}\right) \\ &\quad \times (2.5 \text{ cm}) \left(\frac{1 \text{ kPa}}{10.3 \text{ cm of water}}\right) \\ &= 1.618 \text{ kW} \end{aligned}$$

$$\begin{aligned} \eta &= \frac{P_{\text{ideal}}}{P_{\text{actual}}} \\ &= \left(\frac{1.645 \text{ kW}}{2.7 \text{ kW}}\right) \times 100\% \\ &= 59.93\% \quad (60\%) \end{aligned}$$

The answer is C.

17.

$$\frac{P_2}{P_1} = \left(\frac{N_2}{N_1}\right)^3$$

Assuming the same efficiency at the two speeds,

$$P_2 = P_1 \left(\frac{N_2}{N_1}\right)^3 = (2.7 \text{ kW}) \left(\frac{300 \, \frac{\text{rev}}{\text{min}}}{260 \, \frac{\text{rev}}{\text{min}}}\right)^3$$

$$= 4.148 \text{ kW} \quad (4.2 \text{ kW})$$

The answer is B.

18. The energy equation is

$$\frac{w}{g} + \frac{\text{v}_1^2 - \text{v}_2^2}{2g} + \frac{p_1 - p_2}{\rho g} + z_1 - z_2 = 0$$

But $\text{v}_1 = \text{v}_2$ and $z_1 = z_2$, so that

$$w = \frac{p_2 - p_1}{\rho}$$

$$\begin{aligned} P &= \frac{\dot{m} w}{\eta} \\ &= \left(\frac{\rho Q}{\eta}\right) \left(\frac{p_2 - p_1}{\rho}\right) = \frac{Q(p_2 - p_1)}{\eta} \\ &= \frac{\left(0.015 \, \frac{\text{m}^3}{\text{s}}\right) (400 \text{ kPa})}{0.8} \\ &= 7.5 \text{ kW} \quad (8 \text{ kW}) \end{aligned}$$

The answer is B.

19. The energy equation for a steady state is

$$w + \left(\frac{p}{\rho} + \frac{\text{v}^2}{2} + gz\right)_i = \left(\frac{p}{\rho} + \frac{\text{v}^2}{2} + gz\right)_e$$

$\text{v}_i = \text{v}_e$, $z_i = 0$, and $z_e = 2$ m, so that

$$\begin{aligned} \rho_{\text{oil}} &= (\text{SG})_{\text{oil}} \rho_{\text{water}} \\ &= (0.85) \left(1000 \, \frac{\text{kg}}{\text{m}^3}\right) \\ &= 850 \text{ kg/m}^3 \end{aligned}$$

$$\begin{aligned} w &= \frac{p_e - p_i}{\rho} + g z_e \\ &= \frac{300 \text{ kPa} - (-30 \text{ kPa})}{850 \, \frac{\text{kg}}{\text{m}^3}} \\ &\quad + \left(9.81 \, \frac{\text{m}}{\text{s}^2}\right) (2 \text{ m}) \left(\frac{1 \text{ kJ}}{1000 \text{ J}}\right) \\ &= 0.4079 \text{ kJ/kg} \end{aligned}$$

The mass flow rate is

$$\begin{aligned} \dot{m} &= Q \rho_{\text{oil}} \\ &= \left(0.015 \, \frac{\text{m}^3}{\text{s}}\right) \left(850 \, \frac{\text{kg}}{\text{m}^3}\right) \\ &= 12.75 \text{ kg/s} \end{aligned}$$

The power of the pump is

$$\begin{aligned} P &= \dot{m} w \\ &= \left(12.75 \, \frac{\text{kg}}{\text{s}}\right) \left(0.4079 \, \frac{\text{kJ}}{\text{kg}}\right) \\ &= 5.2 \text{ kW} \quad (5 \text{ kW}) \end{aligned}$$

The answer is D.

20. The force balance on the hovercraft is

$$W = F_{air}$$
$$mg = (\Delta p)A$$
$$\Delta p = \frac{mg}{A} = \frac{(8000\text{ kg})\left(9.81\ \frac{\text{m}}{\text{s}^2}\right)}{(3\text{ m})(8\text{ m})} = 3270\text{ Pa}$$

The velocity of the air across the skirt is given by

$$\frac{v^2}{2} = \frac{\Delta p}{\rho}$$
$$v = \sqrt{\frac{2\Delta p}{\rho}} = \sqrt{\frac{(2)(3270\text{ Pa})}{\left(1.2\ \frac{\text{kg}}{\text{m}^3}\right)}} = 73.82\text{ m/s}$$

$$\dot{m} = \rho A v = \left(1.2\ \frac{\text{kg}}{\text{m}^3}\right)(3\text{ cm})\left(\frac{1\text{ m}}{100\text{ cm}}\right)(2) \times (3\text{ m} + 8\text{ m})\left(73.82\ \frac{\text{m}}{\text{s}}\right) = 58.47\text{ kg/s}$$

$$P_{required} = \dot{m}\left(\frac{\Delta p}{\rho}\right) = \left(58.47\ \frac{\text{kg}}{\text{s}}\right)\left(\frac{(3270\text{ Pa})\left(\frac{1\text{ kPa}}{1000\text{ Pa}}\right)}{1.2\ \frac{\text{kg}}{\text{m}^3}}\right) = 159.33\text{ kW}\quad(160\text{ kW})$$

The answer is C.

21. Since the air jet is open to the atmosphere, the pressure surrounding the jet is constant. Applying the momentum equation in the vertical direction requires equating the weight of the disc to the rate of change of momentum of the air jet, or

$$mg = \rho A v^2$$
$$\rho = \frac{p}{RT} = \frac{101.3\text{ kPa}}{\left(0.287\ \frac{\text{kJ}}{\text{kg·K}}\right)(20°\text{C} + 273°)} = 1.204\text{ kg/m}^3$$

$$v = \sqrt{\frac{mg}{\rho A}} = \sqrt{\frac{(0.01\text{ kg})\left(9.81\ \frac{\text{m}}{\text{s}^2}\right)}{\left(1.204\ \frac{\text{kg}}{\text{m}^3}\right)\left(\frac{\pi}{4}\right)\left((5\text{ mm})\left(\frac{1\text{ m}}{1000\text{ mm}}\right)\right)^2}} = 64.42\text{ m/s}\quad(65\text{ m/s})$$

The answer is B.

22.

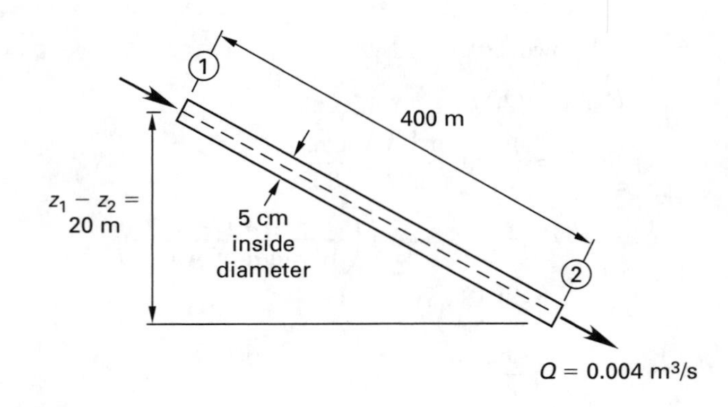

The velocity in the pipe is

$$v = \frac{Q}{A} = \frac{Q}{\left(\frac{\pi}{4}\right)D^2} = \frac{0.004\ \frac{\text{m}^3}{\text{s}}}{\left(\frac{\pi}{4}\right)\left((5\text{ cm})\left(\frac{1\text{ m}}{100\text{ cm}}\right)\right)^2} = 2.036\text{ m/s}$$

The energy equation is

$$\frac{v_1^2 - v_2^2}{2g} + z_1 - z_2 + \frac{p_1 - p_2}{\rho g} = \frac{fLv^2}{2gD}$$

Since $v_1 = v_2$, $z_1 - z_2 = 20$ m, and $p_1 - p_2 = 150$ kPa,

$$20\text{ m} + \frac{(150\text{ kPa})\left(1000\ \frac{\text{Pa}}{\text{kPa}}\right)}{\left(1000\ \frac{\text{kg}}{\text{m}^3}\right)\left(9.81\ \frac{\text{m}}{\text{s}^2}\right)} = \frac{f(400\text{ m})\left(2.036\ \frac{\text{m}}{\text{s}}\right)^2}{(2)\left(9.81\ \frac{\text{m}}{\text{s}^2}\right)\left((5\text{ cm})\left(\frac{1\text{ m}}{100\text{ cm}}\right)\right)}$$

$$f = 0.0209\quad(0.02)$$

The answer is B.

23.

$$\Delta p = \rho g h$$

$$(\text{SG})_{\text{seawater}} = \frac{\rho_{\text{seawater}}}{\rho_{\text{water}}} = 1.03$$

$$\rho_{\text{seawater}} = \rho_{\text{water}} 1.03 = \left(998 \ \frac{\text{kg}}{\text{m}^3}\right)(1.03) = 1027.94 \ \text{kg/m}^3$$

For $h = 100$ m,

$$\Delta p = \rho g h = \left(1027.94 \ \frac{\text{kg}}{\text{m}^3}\right)\left(9.81 \ \frac{\text{m}}{\text{s}^2}\right)(100 \ \text{m}) = 1008.4 \times 10^3 \ \text{Pa} \quad (1000 \ \text{kPa})$$

The answer is B.

24. The velocity at section 1 is

$$\text{v}_1 = \left(\frac{A_2}{A_1}\right)\text{v}_2 = \left(\frac{D_2}{D_1}\right)^2 \text{v}_2 = \left(\frac{10 \ \text{cm}}{20 \ \text{cm}}\right)^2 \left(1 \ \frac{\text{m}}{\text{s}}\right) = 0.25 \ \text{m/s}$$

Apply the Bernoulli equation between sections 1 and 2.

$$\frac{p_1}{\rho} + \frac{\text{v}_1^2}{2} + z_1 = \frac{p_2}{\rho} + \frac{\text{v}_2^2}{2} + z_2$$

$$\frac{(100 \ \text{kPa})\left(1000 \ \frac{\text{Pa}}{\text{kPa}}\right)}{1000 \ \frac{\text{kg}}{\text{m}^3}} + \frac{\left(0.25 \ \frac{\text{m}}{\text{s}}\right)^2}{2} + 0 = \frac{p_2}{1000 \ \frac{\text{kg}}{\text{m}^3}} + \frac{\left(1 \ \frac{\text{m}}{\text{s}}\right)^2}{2} + \left(9.81 \ \frac{\text{m}}{\text{s}^2}\right)(1.5 \ \text{m})$$

Solve for p_2.

$$p_2 = (84816 \ \text{Pa})\left(\frac{1 \ \text{kPa}}{1000 \ \text{Pa}}\right) = 84.82 \ \text{kPa} \quad (80 \ \text{kPa})$$

The answer is B.

25. The Reynolds number is

$$\text{Re} = \frac{\text{v}D\rho}{\mu} = \frac{\left(1.0 \ \frac{\text{m}}{\text{s}}\right)(0.12 \ \text{m})\left(870 \ \frac{\text{kg}}{\text{m}^3}\right)}{0.082 \ \frac{\text{N·s}}{\text{m}^2}} = 1273.17$$

Since Re < 2300, the flow is laminar.

$$f = \frac{64}{\text{Re}} = \frac{64}{1273.17} = 0.05027$$

The head loss is

$$h_f = f\left(\frac{L}{D}\right)\left(\frac{\text{v}^2}{2g}\right) = (0.05027)\left(\frac{100 \ \text{m}}{0.12 \ \text{m}}\right)\left(\frac{\left(1.0 \ \frac{\text{m}}{\text{s}}\right)^2}{(2)\left(9.81 \ \frac{\text{m}}{\text{s}}^2\right)}\right) = 2.135 \ \text{m} \quad (2.1 \ \text{m})$$

The answer is D.

26.

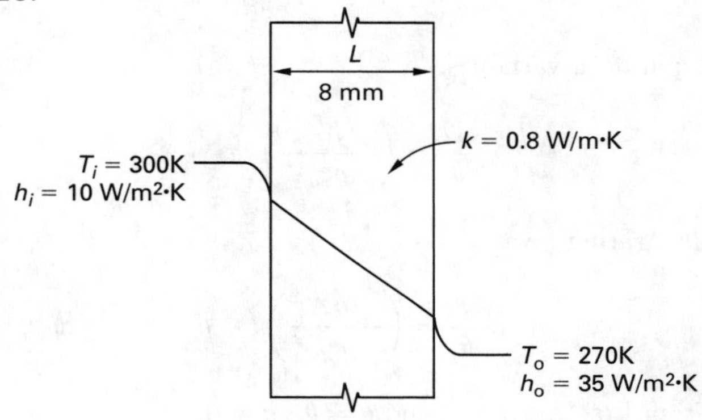

The thermal resistances are

$$R_i = \frac{1}{h_i A} = \frac{1}{\left(10 \ \frac{\text{W}}{\text{m}^2\text{·K}}\right)(1 \ \text{m})(1.5 \ \text{m})} = 6.6667 \times 10^{-2} \ \text{K/W}$$

$$R_{\text{glass}} = \frac{L}{kA} = \frac{(8 \ \text{mm})\left(\frac{1 \ \text{m}}{1000 \ \text{mm}}\right)}{\left(0.8 \ \frac{\text{W}}{\text{m·K}}\right)(1 \ \text{m})(1.5 \ \text{m})} = 6.6667 \times 10^{-3} \ \text{K/W}$$

$$R_o = \frac{1}{h_o A} = \frac{1}{\left(35 \ \frac{\text{W}}{\text{m}^2\text{·K}}\right)(1 \ \text{m})(1.5 \ \text{m})} = 1.9048 \times 10^{-2} \ \text{K/W}$$

The total resistance is

$$R_{\text{total}} = R_i + R_{\text{glass}} + R_o = 6.6667 \times 10^{-2} \ \text{K/W} + 6.6667 \times 10^{-3} \ \text{K/W} + 1.9048 \times 10^{-2} \ \text{K/W} = 9.2382 \times 10^{-2} \ \text{K/W}$$

The heat transfer through the window is

$$Q = \frac{\Delta T}{R_{\text{total}}} = \frac{300\text{K} - 270\text{K}}{9.2382 \times 10^{-2}\ \frac{\text{K}}{\text{W}}}$$
$$= 324.74\ \text{W} \quad (320\ \text{W})$$

The answer is D.

27. The energy balance is

$$hA_s(T - T_\infty) = -\rho c_p V \frac{dT}{dt}$$

Introducing the temperature difference $\theta = T - T_\infty$ gives

$$hA_s\theta = -\rho c_p V \frac{d\theta}{dt}$$

Separating variables,

$$\frac{d\theta}{\theta} = \left(-\frac{hA_s}{\rho c_p V}\right) dt$$

Integrating gives

$$\ln \frac{\theta}{\theta_i} = \left(-\frac{hA_s}{\rho c_p V}\right) t + a \qquad \textit{Eq. 1}$$

From Eq. 1, at $t = 0$ and $\theta = \theta_i$, $a = 0$.

$$\ln \frac{\theta}{\theta_i} = \left(-\frac{hA_s}{\rho c_p V}\right) t$$

Alternatively,

$$\frac{\theta}{\theta_i} = \frac{T - T_\infty}{T_i - T_\infty} = e^{-\left(\frac{hA_s}{\rho c_p V}\right) t} \qquad \textit{Eq. 2}$$

Solve for T from Eq. 2

$$T = T_\infty + (T_i - T_\infty) e^{-\left(\frac{hA_s}{\rho c_p V}\right) t} \qquad \textit{Eq. 3}$$

A portion of the exponent in Eq. 3 is evaluated as

$$\frac{hA_s}{\rho c_p V} = \frac{(400\ \text{W/m}^2\!\cdot\!\text{K})\left(\pi\left((5\ \text{cm})\left(\frac{1\ \text{m}}{100\ \text{cm}}\right)\right)^2\right)}{\left(8933\ \frac{\text{kg}}{\text{m}^3}\right)\left(385\ \frac{\text{J}}{\text{kg}\cdot\text{K}}\right)\left(\frac{\pi}{6}\right) \times \left((5\ \text{cm})\left(\frac{1\ \text{m}}{100\ \text{cm}}\right)\right)^3}$$
$$= 0.01396\ \text{s}^{-1}$$

From Eq. 3

$$T = 300\text{K} + (1000\text{K} - 300\text{K})e^{-(0.01396\ \text{s}^{-1})(30\ \text{s})}$$
$$= 760.5\text{K} \quad (800\text{K})$$

The answer is B.

28.

The heat transfer from the hot oil is

$$Q_H = \dot{m}_H c_H (T_{H_i} - T_{H_o})$$
$$= \left(10\ \frac{\text{kg}}{\text{s}}\right)\left(2\ \frac{\text{kJ}}{\text{kg}\cdot\text{K}}\right)(120°\text{C} - 50°\text{C})$$
$$= 1400\ \text{kW}$$

The heat gain by the cold water is

$$Q_C = \dot{m}_C c_C (T_{C_o} - T_{C_i})$$

Require

$$Q_H = Q_C$$
$$= \dot{m}_C c_C (T_{C_o} - T_{C_i})$$
$$1400\ \text{kW} = \left(8\ \frac{\text{kg}}{\text{s}}\right)\left(4.18\ \frac{\text{kJ}}{\text{kg}\cdot\text{K}}\right)(T_{C_o} - 15°\text{C})$$
$$T_{C_o} = 56.87°\text{C}$$

$$\Delta T_{\text{lm}} = \frac{(T_{H_o} - T_{C_i}) - (T_{H_i} - T_{C_o})}{\ln \frac{T_{H_o} - T_{C_i}}{T_{H_i} - T_{C_o}}}$$
$$= \frac{(50°\text{C} - 15°\text{C}) - (120°\text{C} - 56.87°\text{C})}{\ln \frac{50°\text{C} - 15°\text{C}}{120°\text{C} - 56.87°\text{C}}}$$
$$= 47.69°\text{C} \quad (47.69\text{K})$$

[since it is the difference of temperatures and 1°C = 1K]

$$Q = UA\Delta T_{\text{lm}}$$

Solve for area.

$$A = \left(\frac{Q}{U\Delta T_{\text{lm}}}\right)\left(\frac{1400 \text{ kW}}{\left(0.8 \frac{\text{kW}}{\text{m}^2\cdot\text{K}}\right)(47.69\text{K})}\right)$$
$$= 36.70 \text{ m}^2 \quad (40 \text{ m}^2)$$

The answer is C.

29. The heat transfer to the tank is

$$Q = h_{cr}A\Delta T$$
$$= \left(30 \frac{\text{W}}{\text{m}^2\cdot\text{K}}\right)\pi(1 \text{ m})^2(285\text{K} - 77\text{K})$$
$$= 19\,603.5 \text{ W} \quad (19.60 \text{ kW})$$

The rate of evaporation is

$$\dot{m} = \frac{Q}{h} = \frac{19.60 \text{ kW}}{198 \frac{\text{kJ}}{\text{kg}}}$$
$$= 9.90 \times 10^{-2} \text{ kg/s}$$

The mass of the nitrogen in the tank is

$$m = \rho V$$
$$= \left(810 \frac{\text{kg}}{\text{m}^3}\right)\left(\frac{\pi}{6}\right)(1 \text{ m})^3$$
$$= 424.12 \text{ kg}$$

The time for the nitrogen to be completely vented is

$$t = \frac{m}{\dot{m}} = \frac{424.12 \text{ kg}}{\left(9.90 \times 10^{-2} \frac{\text{kg}}{\text{s}}\right)\left(3600 \frac{\text{s}}{\text{h}}\right)}$$
$$= 1.19 \text{ h} \quad (1 \text{ h})$$

The answer is C.

30.

T_1 = 400K $\quad T_2$ = 300K

$\epsilon_1 \quad \epsilon_2$

$\epsilon_1 = \epsilon_2 = 0.02$

$$Q_{12} = \frac{\sigma(T_1^4 - T_2^4)}{\frac{1-\varepsilon_1}{\varepsilon_1 A_1} + \frac{1}{A_1F_{12}} + \frac{1-\varepsilon_2}{\varepsilon_2 A_2}}$$

Assume $A_1 \approx A_2$ and $F_{12} = 1.0$.

$$q'' = \frac{Q_{12}}{A_1} = \frac{\sigma(T_1^4 - T_2^4)}{\frac{1}{\varepsilon_1} + \frac{1}{\varepsilon_2} - 1}$$
$$= \frac{\left(5.67 \times 10^{-8} \frac{\text{W}}{\text{m}^2\cdot\text{K}^4}\right)\left((400\text{K})^4 - (300\text{K})^4\right)}{\frac{1}{0.02} + \frac{1}{0.02} - 1}$$
$$= 10.023 \text{ W/m}^2 \quad (10 \text{ W/m}^2)$$

The answer is D.

31. The corrosion mass loss is

$$m = \frac{It(\text{MW})}{n\mathcal{F}} \qquad \textit{Eq. 1}$$

The mass of lost steel of thickness d is

$$m = \rho A d \qquad \textit{Eq. 2}$$

Using Eqs. 1 and 2, solve for d.

$$d = \left(\frac{I}{A}\right)\left(\frac{t(\text{MW})}{n\mathcal{F}}\right)\left(\frac{1}{\rho}\right)$$
$$= \left(6 \times 10^{-5} \frac{\text{A}}{\text{cm}^2}\right)$$
$$\times \left(\frac{(4 \text{ yr})\left(365 \frac{\text{d}}{\text{yr}}\right)\left(24 \frac{\text{h}}{\text{d}}\right) \times \left(3600 \frac{\text{s}}{\text{h}}\right)\left(55.85 \frac{\text{g}}{\text{mol}}\right)}{(2)\left(96\,500 \frac{\text{A}\cdot\text{s}}{\text{mol}}\right)\left(7.87 \frac{\text{g}}{\text{cm}^3}\right)}\right)$$
$$= 0.28 \text{ cm}$$

The change in diameter of the steel sphere is

$$\Delta_D = 2d = (2)(0.28 \text{ cm})$$
$$= 0.56 \text{ cm} \quad (0.6 \text{ cm})$$

The answer is B.

32.

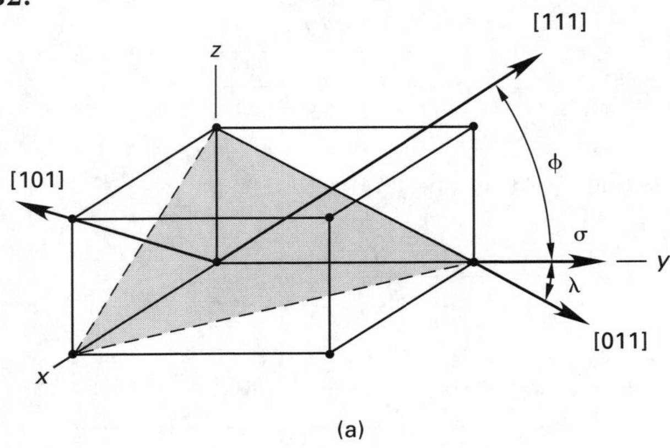

(a)

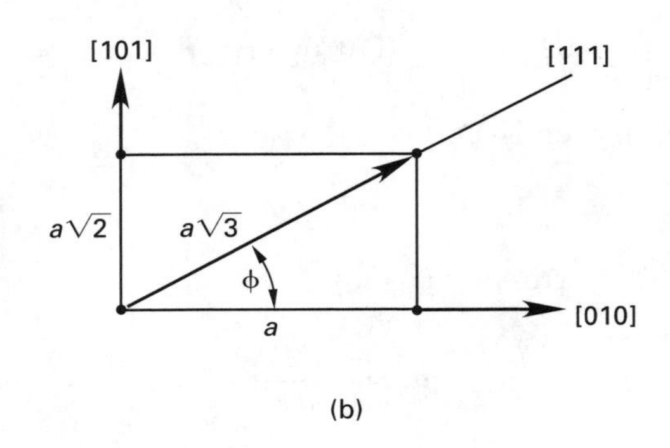

(b)

$$\cos\phi = \frac{a}{a\sqrt{3}}$$
$$\phi = 54.74°$$

$$\begin{aligned}\tau_r &= \sigma\cos\lambda\cos\phi \\ &= (16\text{ MPa})\cos 45°\cos 54.74° \\ &= 6.53\text{ MPa}\quad(6.5\text{ MPa})\end{aligned}$$

The answer is A.

33.

vinyl acetate = $\left[\text{CH}_2\text{–CH(O–C(=O)–CH}_3) \right]_n$

vinyl chloride = $\left[\text{CH}_2\text{–CHCl} \right]_n$

$$\begin{aligned}\text{MW}_{\text{ave}} &= \text{MW}_{\text{VC}}x_{\text{VC}} + \text{MW}_{\text{VA}}x_{\text{VA}} \\ &= \text{MW}_{\text{VC}}x_{\text{VC}} + \text{MW}_{\text{VA}}(1 - x_{\text{VC}}) \quad \textit{Eq. 1}\end{aligned}$$

The average molar mass per polymer is

$$\begin{aligned}\text{MW}_{\text{ave}} &= \frac{\text{MW}_{\text{copolymer}}}{\text{DP}} \\ &= \frac{15\,000\ \frac{\text{g}}{\text{mol}}}{200} \\ &= 75\text{ g/mol}\end{aligned}$$

From Eq. 1

$$75\ \frac{\text{g}}{\text{mol}} = \left(62.5\ \frac{\text{g}}{\text{mol}}\right)x_{\text{VC}} + \left(86\ \frac{\text{g}}{\text{mol}}\right)(1 - x_{\text{VC}})$$
$$x_{\text{VC}} = 0.47\quad(0.5)$$

The answer is C.

34. Two types of austenite are formed during cooling: proeutectic and eutectic.

Using the phase diagram,

% carbon by weight in the liquid at 1148°C = 4.3%
% carbon by weight in the austenite at 1148°C = 2.0%
% carbon by weight in the Fe_3C at 1000°C = 6.67%
% carbon by weight in the austenite at 1000°C = 1.7%

Using these values, the percentage of proeutectic austenite at 1148°C is

$$\begin{aligned}&\frac{\%\text{ C (in liquid)} - \%\text{ C (in original liquid)}}{\%\text{ C (in liquid)} - \%\text{ C (in austenite)}} \\ &\quad= \left(\frac{4.3\% - 3.0\%}{4.3\% - 2.0\%}\right) \times 100\% \\ &\quad= 56.5\%\end{aligned}$$

The total percentage of austenite at 1000°C is

$$\begin{aligned}&\frac{\%\text{ C (in Fe}_3\text{C)} - \%\text{ C (in original liquid)}}{\%\text{ C (in Fe}_3\text{C)} - \%\text{ C (in austenite)}} \\ &\quad= \left(\frac{6.67\% - 3.0\%}{6.67\% - 1.7\%}\right) \times 100\% \\ &\quad= 73.8\%\end{aligned}$$

The percentage of eutectic austenite by weight is

$$73.8\% - 56.5\% = 17.3\%\quad(17\%)$$

The answer is B.

35.

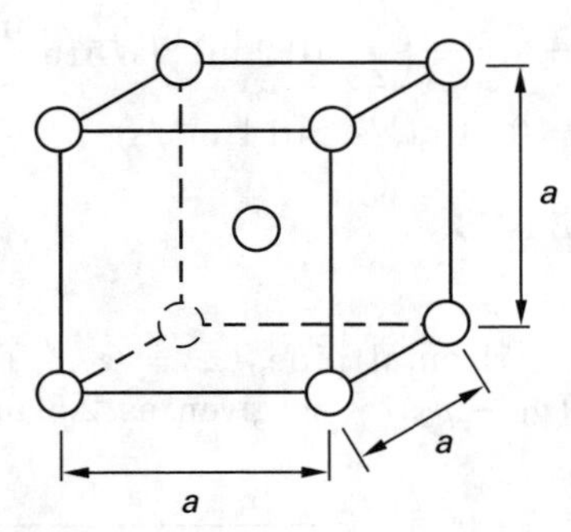

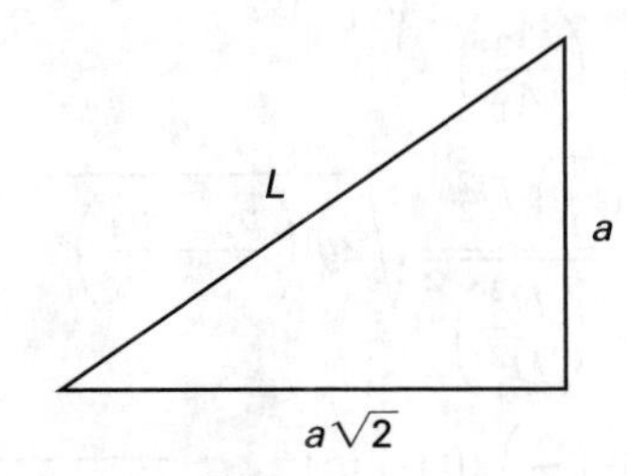

The number of atoms per cell, N, is

$$N = (8)\left(\frac{1}{8}\right) + 1 = 2$$

The volume of an atom is

$$V = \tfrac{4}{3}\pi r^3$$

$$\text{APF} = \frac{V_{\text{atoms/cell}}}{V_{\text{cell}}} = \frac{2\left(\frac{4}{3}\pi r^3\right)}{a^3} \quad \textit{Eq. 1}$$

Along a diagonal, two corner atoms and one central atom are in contact. The diagonal length, L, is

$$\begin{aligned} L &= r + 2r + r = 4r \\ L^2 &= 2a^2 + a^2 \\ L &= \sqrt{3a^2} = 4r \\ a &= \frac{4r}{\sqrt{3}} \end{aligned}$$

Using Eq. 1,

$$\text{APF} = \frac{\frac{8}{3}\pi r^3}{\left(\frac{4}{\sqrt{3}}\right)^3 r^3} = 0.68$$

The answer is B.

36. For laminar flow, the flow rate is given by

$$Q = \frac{\pi D^4 \Delta p_f N}{128\mu L}$$

N is the number of tubes.

At 20°C and 1 atm,

$$\rho = \frac{p}{RT} = \frac{101.325 \text{ kPa}}{\left(0.287 \frac{\text{kJ}}{\text{kg·K}}\right)(20°\text{C} + 273°)} = 1.204 \text{ kg/m}^3$$

From tables, $\mu = 1.81 \times 10^{-8}$ kPa·s.

The flow rate of 20°C atmospheric air is

$$\begin{aligned} Q &= \frac{\pi D^4 \Delta p_f N}{128\mu L} \\ &= \frac{\pi(0.002 \text{ m})^4(0.5 \text{ kPa})(4000)}{(128)(1.81 \times 10^{-8} \text{ kPa·s})(0.25 \text{ m})} \\ &= 0.174 \text{ m}^3/\text{s} \quad (0.2 \text{ m}^3/\text{s}) \end{aligned}$$

To confirm that the flow is laminar, calculate the Reynolds number.

$$\begin{aligned} \text{v} &= \frac{Q}{A} = \frac{Q}{N\left(\frac{\pi}{4}\right)D_{\text{tube}}^2} \\ &= \frac{0.174 \frac{\text{m}^3}{\text{s}}}{(4000)\left(\frac{\pi}{4}\right)\left((2 \text{ mm})\left(\frac{1 \text{ m}}{1000 \text{ mm}}\right)\right)^2} \\ &= 13.813 \text{ m/s} \end{aligned}$$

$$\begin{aligned} \text{Re} &= \frac{\text{v}D\rho}{\mu} \\ &= \frac{\left(13.813 \frac{\text{m}}{\text{s}}\right)\left((2 \text{ mm})\left(\frac{1 \text{ m}}{1000 \text{ mm}}\right)\right) \times \left(1.204 \frac{\text{kg}}{\text{m}^3}\right)}{1.81 \times 10^{-5} \text{ Pa·s}} \\ &= 1837.6 \end{aligned}$$

Since Re < 2300, the flow is laminar and the device is suitable to measure the flow.

Therefore the calculated flow rate of air, $Q = 0.2$ m^3/s, is correct.

The answer is B.

37. The convection heat transfer from the gas to the thermocouple is equal to the radiation heat transfer from the thermocouple to the duct wall.

$$h(T_{\text{gas}} - T_s) = \varepsilon\sigma(T_s^4 - T_w^4)$$

The radiation error is

$$\begin{aligned} T_{\text{gas}} - T_s &= \left(\frac{\varepsilon\sigma}{h}\right)(T_s^4 - T_w^4) \\ &= \left(\frac{0.1}{90\ \frac{\text{W}}{\text{m}^2\cdot\text{K}}}\right)\left(5.679\times10^{-8}\ \frac{\text{W}}{\text{m}^2\cdot\text{K}^4}\right) \\ &\quad\times\left(\begin{array}{l}(800^\circ\text{C}+273^\circ)^4 \\ \quad -(400^\circ\text{C}+273^\circ)^4\end{array}\right) \\ &= 70.7\text{K}\quad(70^\circ\text{C}) \end{aligned}$$

$$\left[\begin{array}{l}\text{since it is the difference} \\ \text{of temperatures and } 1^\circ\text{C} = 1\text{K}\end{array}\right]$$

If a radiation shield is used, this error can be reduced.

The answer is D.

38.

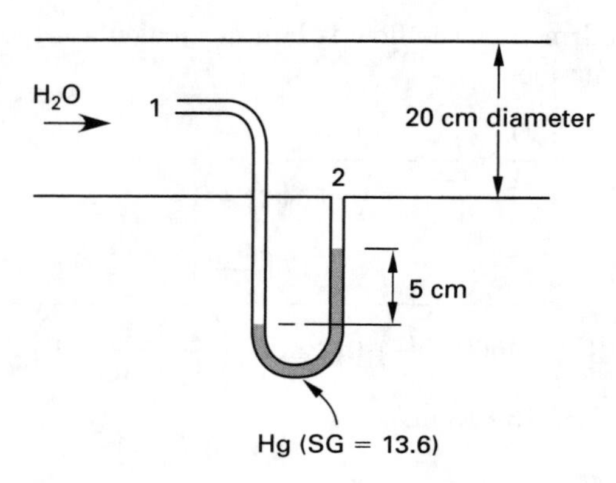

Assuming no head loss between sections 1 and 2,

$$\frac{p_1}{\rho_W} + \frac{\text{v}_1^2}{2} + gz_1 = \frac{p_2}{\rho_W} + \frac{\text{v}_2^2}{2} + gz_2 \qquad \textit{Eq. 1}$$

From Eq. 1 using $\text{v}_1 = 0$ and $z_1 = z_2 = 0$,

$$\begin{aligned} \text{v}_2^2 &= \frac{2(p_1 - p_2)}{\rho_W} \\ &= \frac{2(\rho_{\text{Hg}} - \rho_{\text{H}_2\text{O}})g\Delta h}{\rho_W} \\ &= 2(\text{SG}_{\text{Hg}} - \text{SG}_W)g\Delta h \\ &= (2)(13.6-1)\left(9.81\ \frac{\text{m}}{\text{s}^2}\right)(5\ \text{cm})\left(\frac{1\ \text{m}}{100\ \text{cm}}\right) \\ &= 12.361\ \text{m}^2/\text{s}^2 \\ \text{v}_2 &= 3.516\ \text{m/s} \end{aligned}$$

The flow rate is

$$\begin{aligned} Q = A\text{v} &= \left(\frac{\pi}{4}\right)(0.2\ \text{m})^2\left(3.516\ \frac{\text{m}}{\text{s}}\right) \\ &= 0.11\ \text{m}^3/\text{s}\quad(0.1\ \text{m}^3/\text{s}) \end{aligned}$$

The answer is B.

39. For the horizontal pipe, $z_1 = z_2$. The pressure differential, $(p_1 - p_2)/\gamma$, is given as 2.5 m. The flow rate is

$$\begin{aligned} Q &= \frac{C_\text{v}A_2}{\sqrt{1-\left(\frac{A_2}{A_1}\right)^2}}\sqrt{2g\left(\frac{p_1}{\gamma} + z_1 - \frac{p_2}{\gamma} - z_2\right)} \\ &= \frac{C_\text{v}\left(\frac{\pi}{4}\right)D_2^2}{\sqrt{1-\left(\frac{D_2^2}{D_1^2}\right)^2}}\sqrt{2g\left(\frac{p_1-p_2}{\gamma}\right)} \\ &= \frac{(0.98)\left(\frac{\pi}{4}\right)(0.05\ \text{m})^2}{\sqrt{1-\left(\frac{(0.05\ \text{m})^2}{(0.1\ \text{m})^2}\right)^2}}\sqrt{(2)\left(9.81\ \frac{\text{m}}{\text{s}^2}\right)(2.5\ \text{m})} \\ &= 0.019\ \text{m}^3/\text{s}\quad(0.02\ \text{m}^3/\text{s}) \end{aligned}$$

The answer is B.

40. For an ideal op amp,

$$\begin{aligned} v_1 &= v_2 = v_i \\ i_1 &= 0 \end{aligned}$$

The sum of the currents at node 1 should be equal to zero.

$$\frac{v_1 - 0}{R_1} = \frac{v_o - v_1}{R_f} \qquad \textit{Eq. 1}$$

Rearrange Eq. 1.

$$\left(\frac{R_f}{R_1} + 1\right)v_1 = v_o$$

Then for $v_1 = v_i$, the gain v_o/v_i is

$$\begin{aligned} \frac{v_o}{v_i} &= \frac{R_f}{R_1} + 1 \\ &= \frac{500\ \Omega}{10\ \Omega} + 1 \\ &= 51\quad(50) \end{aligned}$$

The answer is A.

41.

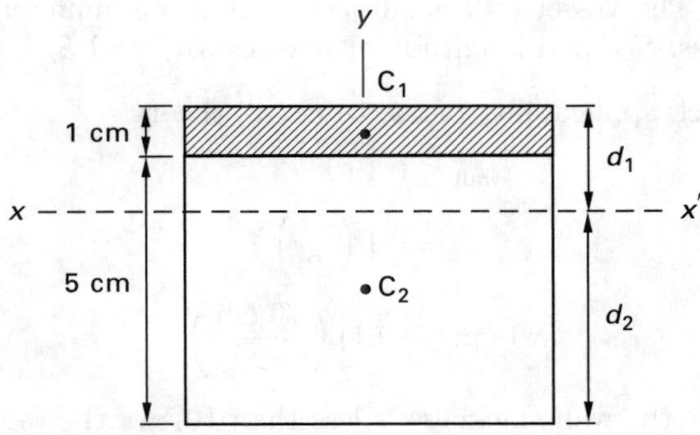

The maximum bending moment will be at the fixed end.

$$M = Fd = (1\ \text{kN})\,(100\ \text{cm})\left(\frac{1\ \text{m}}{100\ \text{cm}}\right) = 1\ \text{kN·m}$$

The y-values of the centroids of the layers in relation to a neutral axis xx' are

$$\begin{aligned} y_{C_1} &= d_1 - \tfrac{1}{2}t_{\text{steel}} \\ y_{C_2} &= d_1 - \left(t_{\text{steel}} + \tfrac{1}{2}t_{\text{wood}}\right) \end{aligned}$$

The neutral axis is located so that

$$\begin{aligned} E_1 y_{C_1} A_1 + E_2 y_{C_2} A_2 &= 0 \\ (195\ \text{GPa})(d_1 - 0.5\ \text{cm})(5\ \text{cm}^2) & \\ +(12\ \text{GPa})(d_1 - 3.5\ \text{cm})(25\ \text{cm}^2) &= 0 \\ d_1 &= 1.21\ \text{cm} \end{aligned}$$

Therefore,

$$\begin{aligned} y_{C_1} &= 1.21\ \text{cm} - 0.5\ \text{cm} = 0.71\ \text{cm} \\ y_{C_2} &= 1.21\ \text{cm} - 3.5\ \text{cm} = -2.29\ \text{cm} \end{aligned}$$

The moment of inertia (about the neutral axis xx') for the steel is

$$\begin{aligned} I_1 &= \frac{b_1 h_1^3}{12} + A_1 y_{C_1}^2 = \frac{(5\ \text{cm})(1\ \text{cm})^3}{12} \\ &\quad + (5\ \text{cm})(1\ \text{cm})(0.71\ \text{cm})^2 \\ &= 2.94\ \text{cm}^4 \quad (2.94 \times 10^{-8}\ \text{m}^4) \end{aligned}$$

Similarly for the wood,

$$\begin{aligned} I_2 &= \frac{b_2 h_2^3}{I^2} + A_z y_{c2}^2 \\ &= \frac{(5\ \text{cm})(5\ \text{cm})^3}{12} + (5\ \text{cm})(5\ \text{cm})(-2.29\ \text{cm})^2 \\ &= 183.2\ \text{cm}^4 \quad (183.2 \times 10^{-8}\ \text{m}^4) \end{aligned}$$

The combined bending rigidity is

$$\begin{aligned} \overline{EI} &= E_1 I_1 + E_2 I_2 \\ &= (195 \times 10^9\ \text{Pa})(2.94 \times 10^{-8}\ \text{m}^4) \\ &\quad + (12 \times 10^9\ \text{Pa})(183.2 \times 10^{-8}\ \text{m}^4) \\ &= 27\,717\ \text{N·m}^2 \quad (27.7\ \text{kN·m}^2) \end{aligned}$$

The maximum stress in steel at $y = d_1$ is

$$\begin{aligned} (\sigma_1)_{\max} &= \frac{ME_1 d_1}{\overline{EI}} \\ &= \frac{(1\ \text{kN·m})(195\ \text{GPa})(1.21\ \text{cm})\left(\dfrac{1\ \text{m}}{100\ \text{cm}}\right)}{27.7\ \text{kN·m}^2} \\ &= (0.085\ \text{GPa})\left(1000\ \frac{\text{MPa}}{\text{GPa}}\right) \\ &= 85\ \text{MPa} \quad (90\ \text{MPa}) \end{aligned}$$

The answer is B.

42. Critical speed in cycles per second for a shaft on two supports is

$$f = \frac{1}{2\pi}\sqrt{\frac{g(m_A\delta_A + m_B\delta_B + \cdots)}{m_A\delta_A^2 + m_B\delta_B^2 + \cdots}}$$

Deflections can be calculated from

$$\begin{aligned} \delta_A &= \left(\frac{Pbx}{6LEI}\right)(L^2 - b^2 - x^2) \\ I &= \left(\frac{\pi}{64}\right)D^4 \end{aligned}$$

The deflection at A due to the 100 kg disk is

$$\begin{aligned} \delta_A &= \left(\frac{\begin{array}{c}(100\ \text{kg})\left(9.81\ \dfrac{\text{m}}{\text{s}^2}\right)(100\ \text{cm}) \\ \times\left(\dfrac{1\ \text{m}}{100\ \text{cm}}\right)(30\ \text{cm})\left(\dfrac{1\ \text{m}}{100\ \text{cm}}\right)\end{array}}{\begin{array}{c}(6)(130\ \text{cm})\left(\dfrac{1\ \text{m}}{100\ \text{cm}}\right)(200\ \text{GPa}) \\ \times\left(10^9\ \dfrac{\text{Pa}}{\text{GPa}}\right)\left(\dfrac{\pi}{64}\right)D^4\end{array}}\right) \\ &\quad \times \left(\begin{array}{c}\left((130\ \text{cm})\left(\dfrac{1\ \text{m}}{100\ \text{cm}}\right)\right)^2 \\ -\left((100\ \text{cm})\left(\dfrac{1\ \text{m}}{100\ \text{cm}}\right)\right)^2 \\ -\left((30\ \text{cm})\left(\dfrac{1\ \text{m}}{100\ \text{cm}}\right)\right)^2\end{array}\right) \\ &= 2.3 \times 10^{-9}\ D^{-4}\ \text{m} \end{aligned}$$

Since only one rotating mass is assumed,

$$\begin{aligned} f &= \frac{1}{2\pi}\sqrt{\frac{gm_A\delta_A}{m_A\delta_A^2}} = \frac{1}{2\pi}\sqrt{\frac{g}{\delta_A}} \\ \omega &= \left(60\frac{\text{s}}{\text{min}}\right)\sqrt{\frac{g}{\delta_A}} \end{aligned}$$

$$500 \text{ rpm} = \left(60 \ \frac{\text{s}}{\text{min}}\right) \sqrt{\frac{9.81 \ \frac{\text{m}}{\text{s}^2}}{2.3 \times 10^{-9} \ D^{-4} \ \text{m}}}$$
$$D = 0.0113 \text{ m} \quad (10 \text{ mm})$$

The answer is B.

43.
$$P = \frac{2\pi TN}{60} \qquad \textit{Eq. 1}$$

Solve for torque from Eq. 1

$$T_{\text{N}\cdot\text{m}} = \frac{60P}{2\pi N} = \frac{(60)(35\,000 \text{ W})}{2\pi N} \qquad \textit{Eq. 2}$$

The maximum allowable torque is derived from

$$T_{\max} = \frac{SJ}{c} \qquad \textit{Eq. 3}$$

Using Eq. 3,

$$T_{\max} = \frac{(70 \text{ MPa})\left(10^6 \ \frac{\text{Pa}}{\text{MPa}}\right)\left(\frac{\pi}{32}\right) \times \left((30 \text{ mm})\left(\frac{1 \text{ m}}{1000 \text{ mm}}\right)\right)^4}{\frac{(30 \text{ mm})\left(\frac{1 \text{ m}}{1000 \text{ mm}}\right)}{2}}$$
$$= 371.1 \text{ N}\cdot\text{m}$$

Using Eq. 2,

$$N = \frac{60P}{2\pi T} = \frac{(60)(35\,000 \text{ W})}{2\pi(371.1 \text{ N}\cdot\text{m})}$$
$$= 900 \text{ rpm}$$

The angle of twist requirement is

$$\theta = \frac{TL}{JG}$$

$$(1.5^\circ)\left(\frac{\pi \text{rad}}{180^\circ}\right) = \frac{T(1 \text{ m})}{\left(\frac{\pi}{32}\right)\left((30 \text{ mm})\left(\frac{1 \text{ m}}{1000 \text{ mm}}\right)\right)^4 \times (75 \text{ GPa})\left(10^9 \ \frac{\text{Pa}}{\text{GPa}}\right)}$$

$$T = 156.1 \text{ N}\cdot\text{m}$$

Using Eq. 2

$$N = \frac{60P}{2\pi T} = \frac{(60)(35\,000 \text{ W})}{2\pi(156.1 \text{ N}\cdot\text{m})}$$
$$= 2142 \text{ rpm}$$

Thus, the maximum speed is dictated by the maximum shear requirement

$$N_{\max} = 900 \text{ rpm}$$

The answer is B.

44. The vessel wall is subject to both circumferential stress, S_{t_c}, and longitudinal stresses, S_{t_L} and S_{t_F}.

Check applicability of the thin-wall theory.

$$t_{\text{wall}} \le 0.1r$$
$$= 0.1\left(\frac{D}{2}\right)$$
$$1 \text{ cm} \le (0.1)\left(\frac{70 \text{ cm}}{2}\right)$$

Since the wall thickness is less than 10% of the radius, the thin-wall approximation is applicable.

Circumferential stress due to the internal pressure is

$$S_{t_c} = \frac{pD_i}{2t}$$
$$= \frac{(1.5 \text{ MPa})\left(10^6 \ \frac{\text{Pa}}{\text{MPa}}\right)(70 \text{ cm})\left(\frac{1 \text{ m}}{100 \text{ cm}}\right)}{(2)(1 \text{ cm})\left(\frac{1 \text{ m}}{100 \text{ cm}}\right)}$$
$$= 52.5 \text{ MPa}$$

Longitudinal stress is

$$S_{t_L} = \frac{pD_i}{4t} + \frac{F}{\pi D_i t}$$
$$= \frac{(1.5 \text{ MPa})(70 \text{ cm})\left(\frac{1 \text{ m}}{100 \text{ cm}}\right)}{(4)(1 \text{ cm})\left(\frac{1 \text{ m}}{100 \text{ cm}}\right)} + \frac{F}{\pi(70 \text{ cm})\left(\frac{1 \text{ m}}{100 \text{ cm}}\right)(1 \text{ cm})\left(\frac{1 \text{ m}}{100 \text{ cm}}\right)}$$
$$= 26.25 \text{ MPa} + 45.5F \qquad \textit{Eq. 1}$$

Then, since $S_{t_c} = 52.5 \text{ MPa} < S_{t_{\max}} = 105 \text{ MPa}$, the vessel can handle the circumferential stress.

Find F from Eq. 1 using the allowable tensile stress.

$$S_{t_{\max}} = S_{t_L} = 105 \text{ MPa}$$
$$= 26.25 \text{ MPa} + 45.5F$$
$$F = 1.73 \text{ MN} \quad (1.7 \text{ MN})$$

The answer is A.

45. All kinetic energy of the object is transformed into strain energy.

$$\text{KE} = \tfrac{1}{2}mv^2 \qquad \textit{Eq. 1}$$
$$\text{SE} = AL\int \sigma d\varepsilon \qquad \textit{Eq. 2}$$
$$\sigma = E\varepsilon \qquad \textit{Eq. 3}$$

Using Eq. 3, Eq. 2 becomes

$$\text{SE} = \frac{AL\sigma^2}{2E} \qquad \textit{Eq. 4}$$

Equate Eqs. 1 and 4, and solve for σ.

$$\sigma = \sqrt{\left(\frac{Em}{AL}\right) \text{v}^2}$$

$$(10 \text{ MPa})\left(10^6 \ \frac{\text{Pa}}{\text{MPa}}\right) = \sqrt{\left(\frac{(17 \text{ MPa})\left(10^6 \ \frac{\text{Pa}}{\text{MPa}}\right) \times (20 \text{ kg})}{\left(\frac{\pi}{4}\right) D^2 (10 \text{ cm}) \times \left(\frac{1 \text{ m}}{100 \text{ cm}}\right)}\right)\left(20 \ \frac{\text{m}}{\text{s}}\right)^2}$$

$$D = 0.132 \text{ m} \quad (13 \text{ cm})$$

The answer is C.

46.

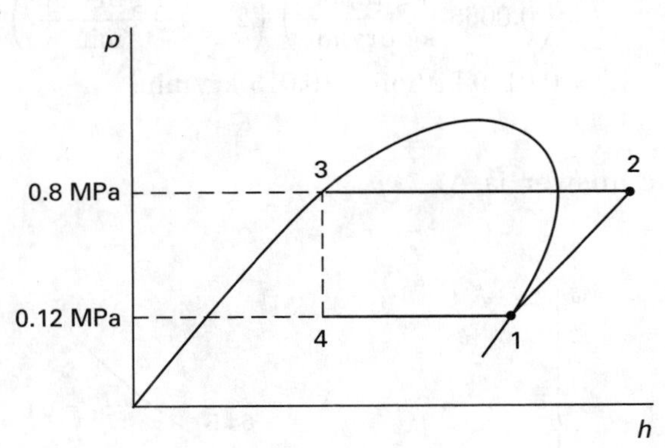

Using refrigerant 134a tables, the properties at the corner points of the cycle are

$$h_1 = 385.4 \text{ kJ/kg}$$
$$s_1 = s_2 = 1.7443 \text{ kJ/kg·K}$$

At 0.8 MPa and $s_2 = 1.7443$ kJ/kg·K,

$$h_2 = 424.8 \text{ kJ/kg}$$
$$T_2 = 40°\text{C}$$
$$h_3 = h_4 = 243.7 \text{ kJ/kg}$$

$$\text{COP} = \frac{h_1 - h_4}{h_2 - h_1} = \frac{385.4 \ \frac{\text{kJ}}{\text{kg}} - 243.7 \ \frac{\text{kJ}}{\text{kg}}}{424.8 \ \frac{\text{kJ}}{\text{kg}} - 385.4 \ \frac{\text{kJ}}{\text{kg}}} = 3.59 \quad (3.6)$$

The answer is B.

47.

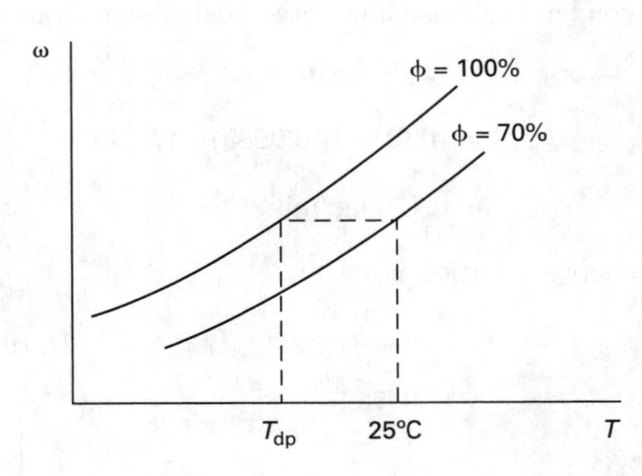

$$\phi = \frac{p_v}{p_{v_{\text{sat}}}}$$

From steam tables at 25°C, $p_{v_{\text{sat}}} = 3.169$ kPa.

$$p_v = \phi p_{v_{\text{sat}}} = (0.7)(3.169 \text{ kPa}) = 2.218 \text{ kPa}$$

At the dew point, $p_v = p_{v_{\text{sat}}} = 2.218$ kPa.

From steam tables, this corresponds to $T_{\text{dp}} \approx 18.9°\text{C}$ (19°C).

The answer is D.

48.

cooling coil
1
2
T_1 = 28°C
ϕ_1 = 0.8
$\dot{V}_1$ = 15 m³/min
T_2 = 12°C
ϕ_2 = 100%
3
$\omega_1 - \omega_2$
T_3 = 12°C

Using the psychrometric chart, the following properties are determined.

$$\omega_1 = 0.0188 \text{ kg of } H_2O\text{/kg of dry air}$$
$$\omega_2 = 0.0085 \text{ kg of } H_2O\text{/kg of dry air}$$
$$v_1 = 0.88 \text{ m}^3\text{/kg dry air}$$
$$h_1 = 76.5 \text{ kJ/kg dry air}$$
$$h_2 = 34.2 \text{ kJ/kg dry air}$$

From steam tables at 12°C, $h_{f_2} = 50.41$ kJ/kg.

The mass flow rate of air is

$$\dot{m}_a = \frac{\dot{V}_1}{v_1} = \frac{15\ \frac{\text{m}^3}{\text{min}}}{0.88\ \frac{\text{m}^3}{\text{kg dry air}}} = 17.046\ \text{kg/min}$$

The condensate mass flow rate is calculated from

$$\begin{aligned}\dot{m}_{\text{condensate}} &= (\omega_1 - \omega_2)\dot{m}_a \\ &= (0.0188 - 0.0085)\left(17.046\ \frac{\text{kg}}{\text{min}}\right) \\ &= 0.176\ \text{kg/min}\end{aligned}$$

The energy equation is

$$Q + \dot{m}_a\left(h_1 - h_2 - (\omega_1 - \omega_2)(h_{f_2})\right) = 0$$

$$Q + \left(17.046\ \frac{\text{kg}}{\text{min}}\right)\left(\begin{array}{l} 76.5\ \frac{\text{kJ}}{\text{kg dry air}} \\ -\ 34.2\ \frac{\text{kJ}}{\text{kg dry air}} \\ -\ (0.0188 - 0.0085) \\ \times\left(50.41\ \frac{\text{kJ}}{\text{kg dry air}}\right)\end{array}\right) = 0$$

$$Q = -712.20\ \text{kJ/min} \quad (-700\ \text{kJ/min})$$

The answer is B.

49.

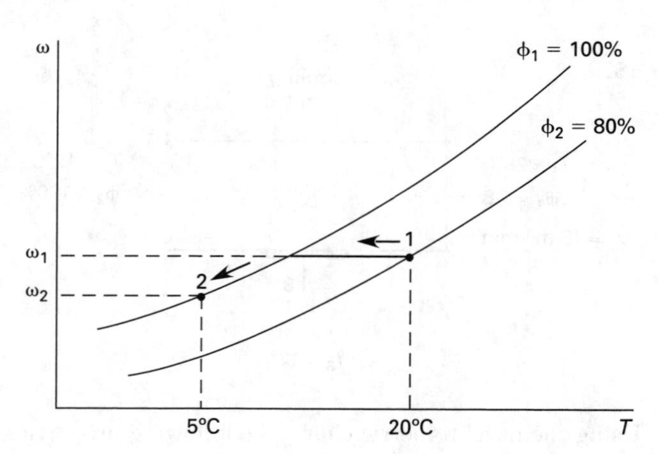

The rate of condensation is

$$\dot{m}_w = (\omega_1 - \omega_2)\dot{m}_a$$

From the psychrometric chart, at 20°C and a relative humidity, ϕ_2, of 80%, $\omega_1 = 0.0117$ kg vapor/kg dry air.

Since 5°C is below the dew point ($\phi_2 = 100\%$),

$$\begin{aligned}\omega_2 &= 0.0054\ \text{kg vapor/kg dry air} \\ \omega_1 - \omega_2 &= 0.0117\ \frac{\text{kg vapor}}{\text{kg dry air}} - 0.0054\ \frac{\text{kg vapor}}{\text{kg dry air}} \\ &= 0.0063\ \text{kg vapor/kg dry air}\end{aligned}$$

From steam tables at 20°C, $p_{v_{\text{sat}}} = 2.339$ kPa.

$$\begin{aligned}p_v &= \phi p_{v_{\text{sat}}} \\ &= (0.8)(2.339\ \text{kPa}) \\ &= 1.8712\ \text{kPa}\end{aligned}$$

The mass flow rate of dry air is

$$\begin{aligned}\dot{m}_a &= \rho_a\dot{V}_a = \left(\frac{p_a}{R_aT}\right)\dot{V}_a \\ &= \left(\frac{101.325\ \text{kPa} - 1.8712\ \text{kPa}}{\left(0.287\ \frac{\text{kJ}}{\text{kg·K}}\right)(20°\text{C} + 273°)}\right)\left(2\ \frac{\text{m}^3}{\text{min}}\right) \\ &= 2.364\ \text{kg/min}\end{aligned}$$

The rate of condensation is

$$\begin{aligned}\dot{m}_\omega &= (\omega_1 - \omega_2)\dot{m}_a \\ &= \left(0.0063\ \frac{\text{kg vapor}}{\text{kg dry air}}\right)\left(2.364\ \frac{\text{kg dry air}}{\text{min}}\right) \\ &= 0.0149\ \text{kg/min} \quad (0.015\ \text{kg/min})\end{aligned}$$

The answer is A.

50.

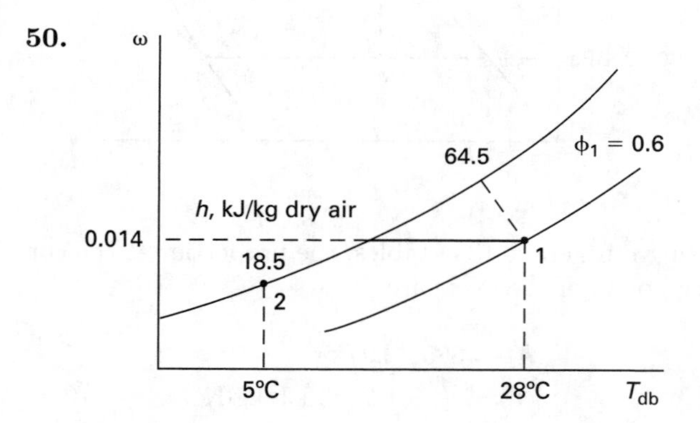

Referring to the psychrometric chart,

$$\begin{aligned}\omega_1 &= 0.014\ \text{kg water vapor/kg dry air} \\ h_1 &= 64.5\ \text{kJ/kg dry air} \\ h_2 &= 18.5\ \text{kJ/kg dry air}\end{aligned}$$

$$m_a = \frac{m}{1+\omega_1} = \frac{5\ \text{kg}}{1+0.014}$$
$$= 4.93\ \text{kg}$$
$$Q_{12} = m_a(h_2 - h_1)$$
$$= (4.93\ \text{kg})\left(18.5\ \frac{\text{kJ}}{\text{kg dry air}} - 64.5\ \frac{\text{kJ}}{\text{kg dry air}}\right)$$
$$= -226.78\ \text{kJ}\quad (-230\ \text{kJ})$$

The answer is B.

51. Due to symmetry, the point C will move down vertically, as shown.

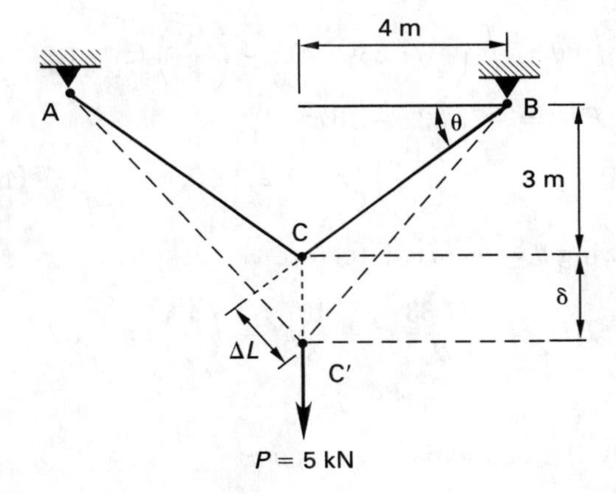

The stress in each member is

$$\sigma_{AC} = E\varepsilon_{AC} = E\left(\frac{\Delta L}{L}\right)_{AC}$$
$$= E\left(\frac{(\sin\theta)\delta}{L_{AC}}\right) \qquad \textit{Eq. 1}$$
$$\sigma_{BC} = E\left(\frac{(\sin\theta)\delta}{L_{BC}}\right) = \sigma_{AC}$$
$$F_{AC} = \sigma_{AC}A_{AC} \qquad \textit{Eq. 2}$$

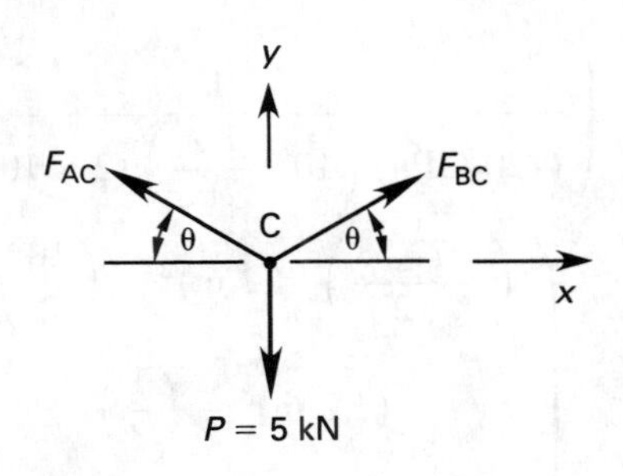

$$\sum F_c\Big|_y = 0$$
$$2F_{AC}\sin\theta = P$$
$$F_{AC} = \frac{P}{2\sin\theta} \qquad \textit{Eq. 3}$$

Using Eqs. 1, 2, and 3, solve for δ.

$$\delta = \frac{PL_{AC}}{2\sin^2\theta(AE)_{AC}}$$
$$= \frac{(5\ \text{kN})\left(1000\ \frac{\text{N}}{\text{kN}}\right)(5\ \text{cm})}{(2)\left(\frac{3}{5}\right)^2(1\ \text{cm}^2)\left(\frac{1\ \text{m}}{100\ \text{cm}}\right)^2 \times (200\ \text{GPa})\left(10^9\ \frac{\text{Pa}}{\text{GPa}}\right)}$$
$$= 0.0017\ \text{m}$$

Using Eq. 1,

$$\sigma_{AC} = E\left(\frac{(\sin\theta)\delta}{L_{AC}}\right)$$
$$= (200\ \text{GPa})\left(10^9\ \frac{\text{Pa}}{\text{GPa}}\right)\left(\frac{\left(\frac{3}{5}\right)(0.0017\ \text{m})}{5\ \text{m}}\right)$$
$$= 40.8\times 10^6\ \text{Pa}\quad (41\ \text{MPa})$$

The answer is C.

52.

$$T_{st} + T_{br} = T = 350\ \text{N·m} \qquad \textit{Eq. 1}$$
$$\phi_{st} = \phi_{br}$$
$$\phi_{st} = \frac{T_{st}L_{st}}{J_{st}G_{st}} \qquad \textit{Eq. 2}$$
$$\phi_{br} = \frac{T_{br}L_{br}}{J_{br}G_{br}} \qquad \textit{Eq. 3}$$
$$J_{br} = \frac{\pi}{2}r_{br}^4 \qquad \textit{Eq. 4}$$
$$J_{st} = \frac{\pi}{2}\left(r_{st}^4 - r_{br}^4\right) \qquad \textit{Eq. 5}$$

Using Eqs. 2, 3, 4, and 5,

$$\frac{T_{st}L_{st}}{\left(\frac{\pi}{2}\right)\left(\left((30\ \text{mm})\left(\frac{1\ \text{m}}{1000\ \text{mm}}\right)\right)^4 - \left((10\ \text{mm})\left(\frac{1\ \text{m}}{1000\ \text{mm}}\right)\right)^4\right) \times (75\ \text{GPa})\left(10^9\ \frac{\text{Pa}}{\text{GPa}}\right)}$$
$$= \frac{T_{br}L_{br}}{\left(\frac{\pi}{2}\right)\left((10\ \text{mm})\left(\frac{1\ \text{m}}{1000\ \text{mm}}\right)\right)^4 \times (35\ \text{GPa})\left(10^9\ \frac{\text{Pa}}{\text{GPa}}\right)}$$

$$T_{st} = 171.43\ T_{br} \qquad \textit{Eq. 6}$$

Using Eqs. 1 and 6,

$$\begin{aligned} T_{st} + T_{br} &= 350\ \text{N·m} \\ 171.43\ T_{br} + T_{br} &= 350\ \text{N·m} \\ T_{br} &= 2.03\ \text{N·m} \end{aligned}$$

Using Eqs. 3 and 4,

$$\begin{aligned} \phi_{br} &= \frac{T_{br}L_{br}}{J_{br}G_{br}} = \frac{T_{br}L_{br}}{\left(\frac{\pi}{2}\right) r_{br}^4 G_{br}} \\ &= \frac{(2.03\ \text{N·m})(1.5\ \text{m})}{\left(\left(\frac{\pi}{2}\right)\left((10\ \text{mm})\left(\frac{1\ \text{m}}{1000\ \text{mm}}\right)\right)^4 \times (35\ \text{GPa})\left(10^9\ \frac{\text{Pa}}{\text{GPa}}\right)\right)} \\ &= 0.0055\left(\frac{180}{\pi}\right) \\ &= 0.3^\circ \end{aligned}$$

The answer is B.

53. Determine R and M_0 for the free-body diagram shown.

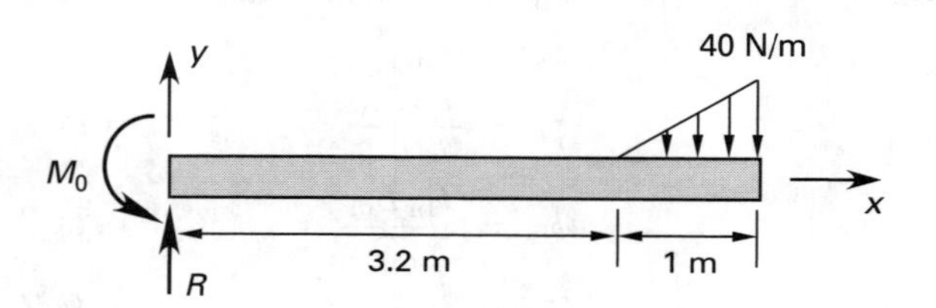

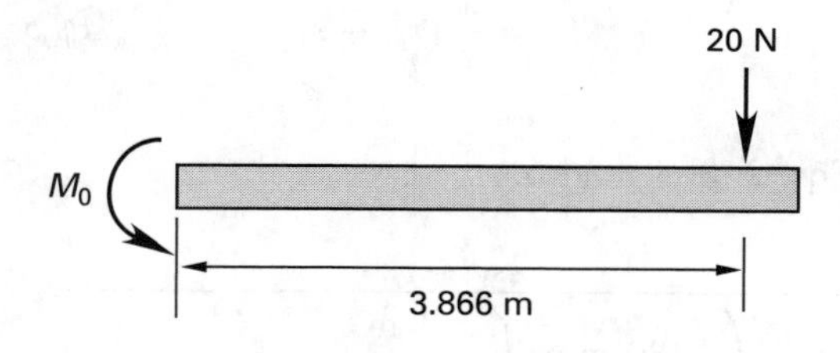

$$\sum F_x = 0$$

$$\sum F_y = 0$$

$$R - \left(\frac{1}{2}\right)(1\ \text{m})\left(40\ \frac{\text{N}}{\text{m}}\right) = 0$$

$$R = 20\ \text{N}$$

$$\sum M_0 = 0$$

$$\begin{aligned} M_0 &= (20\ \text{N})(3.866\ \text{m}) \\ &= 77.33\ \text{N·m} \end{aligned}$$

The moment distribution over the beam in the range $x > 3.2$ m is

$$M = -77.33\ \text{N·m} + (20\ \text{N})x - \left(\frac{40\ \frac{\text{N}}{\text{m}}}{6}\right)(x - 1.2)^3$$

$$EI\left(\frac{d\theta}{dx}\right) = M \qquad \textit{Eq. 1}$$

$$EI\left(\frac{d^2y}{dx^2}\right) = M$$

Integrate Eq. 1.

$$EI\int d\theta = \int\left(-77.33 + 20x - \left(\frac{40}{6}\right)(x - 1.2)^3\right)dx$$

$$EI\theta = -77.33x + 10x^2 - \left(\frac{10}{6}\right)(x - 1.2)^4 + c_1 \qquad \textit{Eq. 2}$$

Knowing $\theta = dy/dx$, integrate Eq. 2.

$$EIy = -\frac{77.33}{2}x^2 + \frac{10}{3}x^3 - \left(\frac{1}{3}\right)(x - 1.2)^5 + c_1x + c_2$$

Apply the boundary conditions.

$$y = 0 \text{ at } x = 0$$
$$\theta = 0 \text{ at } x = 0$$

Thus

$$c_1 \text{ and } c_2 = 0$$

The maximum deflection will be at $x = 4.2$ m.

$$\begin{aligned} y_{max,m} &= \left(\frac{1}{EI}\right)\left(\frac{-77.33}{2}x^2 + \frac{10}{3}x^3 - \left(\frac{1}{3}\right)(x - 1.2)^5\right) \\ &= \left(\frac{1}{(120\ \text{GPa})\left(10^9\ \frac{\text{Pa}}{\text{GPa}}\right)(2\times10^{-6}\ \text{m}^4)}\right) \\ &\quad\times\left(\left(-\frac{77.33}{2}\right)(4.2\ \text{m})^2 + \left(\frac{10}{3}\right)\times(4.2\ \text{m})^3 - \left(\frac{1}{3}\right)\times(4.2\ \text{m} - 1.2\ \text{m})^5\right) \\ &= -0.0028\ \text{m} \quad (2\ \text{mm}) \end{aligned}$$

The minus sign indicates that the maximum deflection is in the downward, or $-y$, direction.

The answer is A.

54. The expansion or elongation of the pipe due to heat is

$$\delta_T = \alpha L(T - T_o) = \left(19 \times 10^{-6} \frac{1}{^\circ\text{C}}\right) \times (1.5 \text{ m})(120^\circ\text{C} - 25^\circ\text{C}) = 2.71 \times 10^{-3} \text{ m}$$

This elongation is countered by compression by the wall, thus producing compressive stress.

$$\sigma = E\varepsilon = E\left(\frac{\delta_T}{L}\right) = (100 \text{ GPa})\left(10^9 \frac{\text{Pa}}{\text{GPa}}\right)\left(\frac{2.71 \times 10^{-3} \text{ m}}{1.5 \text{ m}}\right) = 181 \times 10^6 \text{ Pa} \quad (180 \text{ MPa})$$

The answer is D.

55. The maximum bending moment will be at the wall.

The sum of moments about the neutral axis is

$$\sum M_{NA} = 0$$

$$\left((4 \text{ kN})\sin 75^\circ(2 \text{ m}) + (4 \text{ kN})\cos 75^\circ(9 \text{ mm}) \times \left(\frac{1 \text{ m}}{1000 \text{ mm}}\right) - M_0\right) = 0$$

$$M_0 = 7.74 \text{ kN·m} \quad (7.74 \times 10^3 \text{ N·m})$$

$$\sigma_{\max} = \frac{Mc}{I}$$

$$I = 2(I_{c,\text{ flange}} + Ad^2) + I_{\text{web}} = 2\left(\tfrac{1}{12}b_f h_f^3 + A_f d^2\right) + \tfrac{1}{12}b_w h_w^3$$

$$= (2)\left(\begin{array}{l}\left(\frac{1}{12}\right)(12 \text{ mm})\left(\frac{1 \text{ m}}{1000 \text{ mm}}\right) \times \left((3 \text{ mm})\left(\frac{1 \text{ m}}{1000 \text{ mm}}\right)\right)^3 \\ + (12)\left(\frac{1 \text{ m}}{1000 \text{ mm}}\right) \times (3 \text{ mm})\left(\frac{1 \text{ m}}{1000 \text{ mm}}\right) \\ \times \left(\left(\frac{12 \text{ mm}}{2} + \frac{3 \text{ mm}}{2}\right)\left(\frac{1 \text{ m}}{1000 \text{ mm}}\right)\right)\end{array}\right) + \left(\frac{1}{12}\right)(3 \text{ mm})\left(\frac{1 \text{ m}}{1000 \text{ mm}}\right) \times \left((12)\left(\frac{1 \text{ m}}{1000 \text{ mm}}\right)\right)^3$$

$$= 5.4 \times 10^{-7} \text{ m}^4$$

$$c = \left(\left(\frac{12 \text{ mm}}{2} + \frac{3 \text{ mm}}{2}\right)\left(\frac{1 \text{ m}}{1000 \text{ mm}}\right)\right) = 7.5 \times 10^{-3} \text{ m}$$

$$\sigma_{\max}\big|_{\text{bending}} = \frac{(7.74 \times 10^3 \text{ N·m})(0.0075 \text{ m})}{5.4 \times 10^{-7} \text{ m}^4} = 1.08 \times 10^{10} \text{ Pa}$$

$$\sigma_{\text{normal}} = \frac{(4 \text{ kN})\left(1000 \frac{\text{N}}{\text{kN}}\right)\cos 75^\circ}{\left((3)(12 \text{ mm})\left(\frac{1 \text{ m}}{1000 \text{ mm}}\right) \times (3 \text{ mm})\left(\frac{1 \text{ m}}{1000 \text{ mm}}\right)\right)} = 14.38 \times 10^6 \text{ Pa}$$

$$\sigma_{\max} = 1.08 \times 10^{10} \text{ Pa} + 14.38 \times 10^6 \text{ Pa} = 10.8 \times 10^9 \text{ Pa} \quad (11 \text{ GPa})$$

The answer is A.

56.

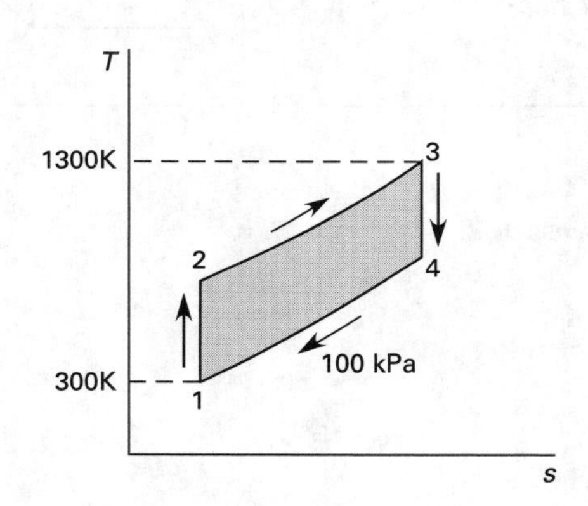

$$\eta_t = \frac{|w_{\text{net}}|}{q_{\text{in}}} = \frac{|w_t - w_c|}{q_{\text{in}}}$$

$$|w_t| = h_3 - h_4 = c_p(T_3 - T_4)$$

$$|w_c| = h_2 - h_1 = c_p(T_2 - T_1)$$

$$q_{\text{in}} = h_3 - h_2 = c_p(T_3 - T_2)$$

Processes 1–2 and 3–4 are isentropic.

$$\frac{T_2}{T_1} = \left(\frac{p_2}{p_1}\right)^{\frac{k-1}{k}} = r_p^{\frac{k-1}{k}}$$

$$T_2 = (300\text{K})10^{\frac{1.36-1}{1.36}} = 551.9\text{K}$$

$$\frac{T_4}{T_3} = \left(\frac{p_2}{p_3}\right)^{\frac{k-1}{k}}$$

$$T_4 = (1300\text{K})\left(\frac{1}{10}\right)^{\frac{1.36-1}{1.36}} = 706.7\text{K}$$

The thermal efficiency is

$$\begin{aligned}\eta &= \frac{c_p(T_3 - T_4) - c_p(T_2 - T_1)}{c_p(T_3 - T_2)} \\ &= \frac{(T_3 - T_4) - (T_2 - T_1)}{T_3 - T_2} \\ &= \frac{(1300\text{K} - 706.7\text{K}) - (551.9\text{K} - 300\text{K})}{1300\text{K} - 551.9\text{K}} \\ &= 0.4564 \quad (46\%)\end{aligned}$$

The answer is D.

57.

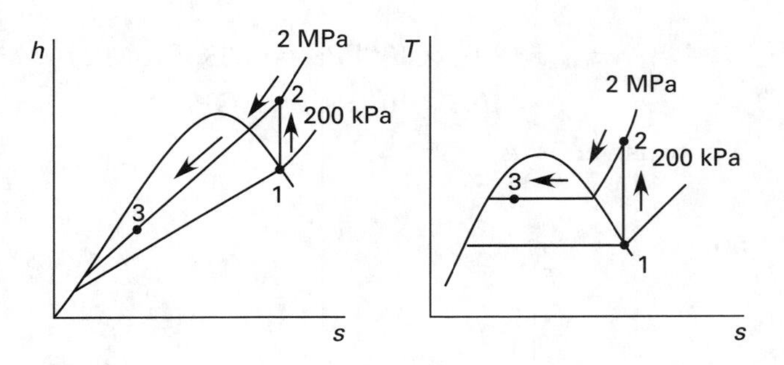

For process 1–2,

$$q_{12} = 0$$

For process 2–3,

$$q_{23} = h_3 - h_2$$

From steam tables,

At 200 kPa, $s_1 = s_2 = s_g = 7.1271$ kJ/kg·K.
At 2 MPa and $s_2 = 7.1271$ kJ/kg, superheated tables give $h_2 = 3247.6$ kJ/kg.
At 2 MPa and $x_3 = 0.3$,

$$\begin{aligned}h_3 &= h_f + x_3 h_{fg} \\ &= 908.79\ \frac{\text{kJ}}{\text{kg}} + (0.3)\left(1890.7\ \frac{\text{kJ}}{\text{kg}}\right) \\ &= 1476\ \text{kJ/kg} \\ q_{23} &= h_3 - h_2 \\ &= 1476\ \frac{\text{kJ}}{\text{kg}} - 3247.6\ \frac{\text{kJ}}{\text{kg}} \\ &= -1771.6\ \text{kJ/kg} \quad (-1800\ \text{kJ/kg})\end{aligned}$$

The answer is A.

58.

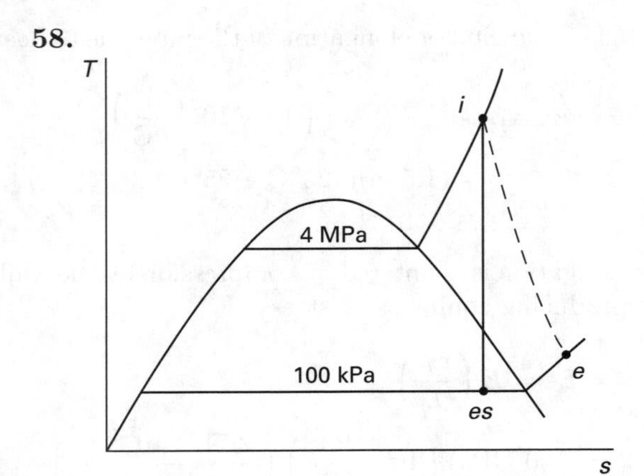

From superheated steam tables at 4 MPa and 400°C,

$$h_i = 3213.6\ \text{kJ/kg}$$
$$s_i = 6.769\ \text{kJ/kg·K}$$

At 100 kPa and 150°C,

$$h_e = 2776.4\ \text{kJ/kg}$$
$$s_e = 7.6134\ \text{kJ/kg·K}$$

The isentropic efficiency of the turbine is

$$\eta_s = \frac{h_i - h_e}{h_i - h_{es}} \qquad \textit{Eq. 1}$$

To determine h_{es},

$$\begin{aligned}s_i &= s_{es} \\ 6.769\ \frac{\text{kJ}}{\text{kg·K}} &= s_f + x_{es} s_{fg} \\ &= 1.30126\ \frac{\text{kJ}}{\text{kg·K}} + x_{es}\left(6.0568\ \frac{\text{kJ}}{\text{kg·K}}\right) \\ x_{es} &= 0.9028 \\ h_{es} &= h_f + x_{es} h_{fg} \\ &= 417.46\ \frac{\text{kJ}}{\text{kg}} + (0.9028)\left(2258\ \frac{\text{kJ}}{\text{kg}}\right) \\ &= 2455.86\ \text{kJ/kg}\end{aligned}$$

Using these enthalpies in Eq. 1,

$$\begin{aligned}\eta_s &= \frac{h_i - h_e}{h_i - h_{es}} \\ &= \frac{3213.6\ \frac{\text{kJ}}{\text{kg}} - 2776.4\ \frac{\text{kJ}}{\text{kg}}}{3213.6\ \frac{\text{kJ}}{\text{kg}} - 2455.86\ \frac{\text{kJ}}{\text{kg}}} \\ &= 0.576 \quad (60\%)\end{aligned}$$

The answer is C.

59.

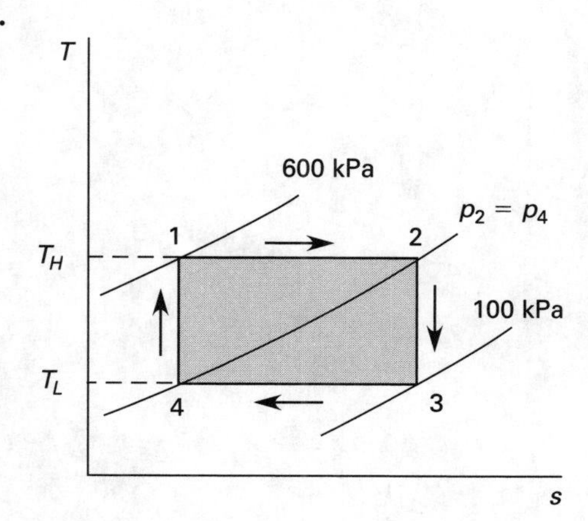

Processes 2–3 and 4–1 are isentropic.

$$k = \frac{c_p}{c_V} = \frac{1.0035\ \frac{\text{kJ}}{\text{kg·K}}}{0.7165\ \frac{\text{kJ}}{\text{kg·K}}} = 1.4$$

$$\frac{T_H}{T_L} = \left(\frac{p_1}{p_4}\right)^{\frac{k-1}{k}} = \left(\frac{p_2}{p_3}\right)^{\frac{k-1}{k}}$$

$$\frac{p_1}{p_4} = \frac{p_2}{p_3}$$

Or

$$\begin{aligned} p_2 = p_4 &= \sqrt{p_1 p_3} \\ &= \sqrt{(600\ \text{kPa})(100\ \text{kPa})} \\ &= 244.95\ \text{kPa} \end{aligned}$$

$$\frac{T_H}{T_L} = \left(\frac{p_1}{p_4}\right)^{\frac{k-1}{k}}$$

$$\frac{T_H}{300\text{K}} = \left(\frac{600\ \text{kPa}}{244.95\ \text{kPa}}\right)^{\frac{1.4-1}{1.4}}$$

$$T_H = 387.507\text{K}$$

$$\begin{aligned} \Delta s &= c_p \ln\frac{T_2}{T_1} - R\ln\frac{p_2}{p_1} \\ &= 0 - \left(0.287\ \frac{\text{kJ}}{\text{kg·K}}\right)\ln\frac{244.95\ \text{kPa}}{600\ \text{kPa}} \\ &= 0.257\ \text{kJ/kg·K} \end{aligned}$$

$$\begin{aligned} w &= \text{area 1-2-3-4-1} = (\Delta T)(\Delta s) \\ &= (387.507\text{K} - 300\text{K})\left(0.257\ \frac{\text{kJ}}{\text{kg·K}}\right) \\ &= 22.49\ \text{kJ/kg} \quad (22\ \text{kJ/kg}) \end{aligned}$$

The answer is B.

60.

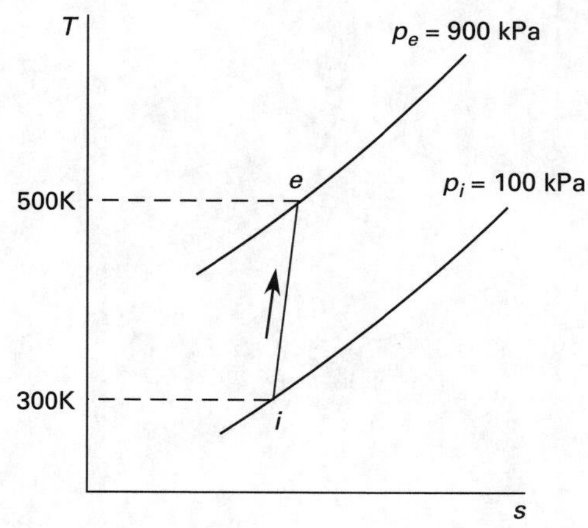

To determine the exponent n,

$$\frac{T_e}{T_i} = \left(\frac{p_e}{p_i}\right)^{\frac{n-1}{n}}$$

$$\frac{500\text{K}}{300\text{K}} = \left(\frac{900\ \text{kPa}}{100\ \text{kPa}}\right)^{\frac{n-1}{n}}$$

$$\ln\frac{5}{3} = \left(\frac{n-1}{n}\right)\ln 9$$

$$n = 1.303$$

The work done per unit mass is

$$\begin{aligned} w_{ie} &= \left(\frac{n}{n-1}\right)(p_e v_e - p_i v_i) \\ &= \left(\frac{n}{n-1}\right) R(T_e - T_i) \\ &= \left(\frac{1.303}{1.303 - 1}\right)\left(0.287\ \frac{\text{kJ}}{\text{kg·K}}\right)(500\text{K} - 300\text{K}) \\ &= 246.8\ \text{kJ/kg} \quad (250\ \text{kJ/kg}) \end{aligned}$$

The answer is B.